The Technology of Offshore Drilling, Completion and Production

The Technology of Offshore Drilling, Completion and Production

Compiled by
ETA Offshore Seminars, Inc.

THE PETROLEUM PUBLISHING COMPANY
Tulsa

Library of Congress Catalog Card Number 75-21903
International Standard Book Number 0-87814-066-2
Printed in U.S.A.

Contents

Foreword

The collection of articles contained herein represents a unique effort by the international training company, ETA Offshore Seminars, Inc., to provide a comprehensive and up-to-date learning tool for the offshore petroleum industry. At a time when a severe shortage of trained rig personnel exists, the need for quick and efficient teaching devices is critical.

The multifaceted direction of this book is designed to accommodate not only the newcomers to the industry but also the veteran personnel. The articles cover both introductory aspects of offshore drilling, completion, and production and further delve into the technical details normally not found in beginner texts. The authors, who represent vital segments of the offshore industry, are leading experts in their fields.

The staff of ETA Offshore Seminars, Inc. gratefully acknowledges the dedication and hard work contributed by each of the authors in the preparation of the articles. The publication of a text of this nature is a tribute both to the contributors and to the industry for which it is written. The compilation of such a vast amount of knowledge and expertise under one cover is a tremendous accomplishment.

I would also like to acknowledge the efforts and long hours spent in the preparation of this volume by Ralph G. McTaggart, Bonnie S. Somyak and Susan Huey also the staff of ETA Offshore Seminars, Inc.

R. Stewart Hall
Chairman of the Board,
ETA Offshore Seminars, Inc.
Houston, Texas
1975

Introduction

The offshore drilling industry is involved in an unprecedented construction and drilling boom. Hundreds of millions of dollars are being spent to capture the much-needed energy supplies. The technology for drilling and producing in deeper waters and in more hostile environments is rapidly expanding. These factors have created an acute shortage of trained and qualified personnel to man the rigs now under construction.

Recognizing this need, ETA Offshore Seminars, Inc. (ETAOS) is attempting to alleviate the problem by presenting seminars aimed at training new men and updating experienced personnel on the theory, selection, and operation of offshore drilling, subsea completion, and production equipment.

The articles presented in this text are derivatives of the presentations made at the ETAOS international training seminars. The faculty of top experts from leading offshore disciplines that has presented numerous ETAOS seminars worldwide has authored the articles for this book.

The contents of the articles range from the very basic descriptions of equipment to comprehensive details of the latest technology in the industry. Illustrative drawings and photographs further supplement the text and allow for easier and more detailed comprehension. The tremendous amount of knowledge contained herein is equivalent to experience gained only through months or years of exposure in the offshore industry.

PART I

Offshore Technology

1

Offshore Mobile Drilling Units

Ralph G. McTaggart
ETA Engineers, Inc.

Offshore mobile drilling units as we know them today are sophisticated pieces of machinery. However, this was not always the case. The original units were simply land rigs taken into shallow waters and placed on a structure for drilling. The same drilling techniques that had been developed on land were used on the first offshore rigs. These techniques worked for some time, but the need to drill in deeper waters created a new type of engineer—the offshore structural design engineer. And along with the new engineering concepts came the new breed of drilling rigs which we see today.

Following drilling trends, we find that there are four basic types of offshore mobile drilling units: the submersible, the jack-up, the semisubmersible, and the drillship. A typical evolution process is shown in Figure 1-1.

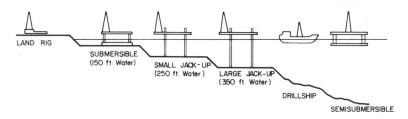

Fig. 1–1 *Typical evolution process for offshore mobile drilling units.*

3

To date, fixed structures have been installed in water depths up to 350 feet, but in this presentation we shall just discuss the movable structures.

Submersible Drilling Rigs

Figure 1-2 illustrates the evolution of the submersible drilling unit, sometimes called the swamp barge or posted barge. This type of unit is used in shallow waters such as rivers and bays, usually in waters up to 50 feet deep. One submersible, however, has been used in 175 foot water depths. The submersible has two hulls. The upper hull, sometimes referred to as the "Texas" deck, is used to house the crew quarters and equipment, and the drilling is performed through a slot on the stern with a cantilevered structure. The lower hull is the ballast area and is also the foundation used while drilling.

The submersible is floated to location like a conventional barge and is then ballasted to rest on the river bottom. The lower hulls are designed to withstand the weight of the total unit and the drilling load.

Stability while ballasting these units is a critical factor. In fact, the techniques developed were the foundation of semisubmersible ballasting schedules. As a point of interest, the first semisubmersibles were converted submersibles. Today, however, submersibles are fading from the scene simply because current water depth requirements have surpassed their capabilities. For example, in 1974 there were only about 25 submersibles in operation.

Jack-up Units

The first "jack-up" rig was the DeLong Rig No. 1, which was built in 1950 but was permanently installed as a platform in 1953. Figure 1-3 shows the evolution of the jack-up. The first "Mobile Jack-up" was the DeLong-McDermott No. 1, later to be known as The Offshore Company Rig No. 51.

This unit was followed by Glasscock Drilling Company's Mr. Gus (Figure 1-4) and by The Offshore Company Rig No. 52 (Figure 1-5). Each of these jack-up rigs had multiple piles or legs.

CHRONOLOGY OF SUBMERSIBLE RIGS, 1949–1968

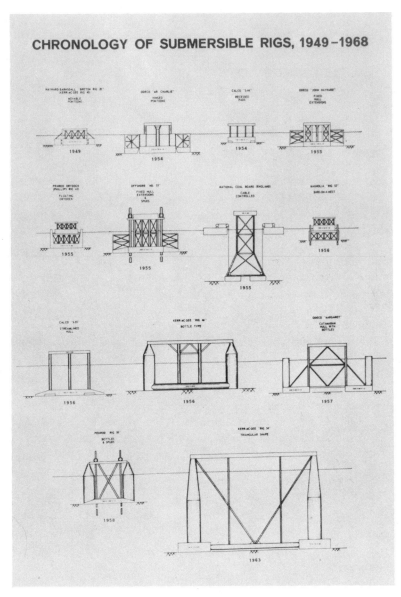

Fig. 1–2 *Submersible evolution.*

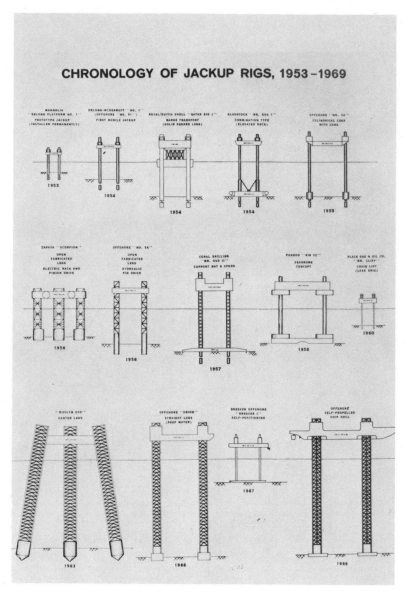

Fig. 1–3   *Jack-up evolution.*

Fig. 1–4 *Glasscock's Mr. Gus I.*

Fig. 1–5 *Offshore's rig 52.*

In 1955, the first 3-legged jack-up appeared on the scene (Figure 1-6). The rig was the R. G. LeTourneau jack-up, the Scorpion, for Zapata Offshore Company. The Scorpion, an independent leg jack-up, used a rack and pinion elevating system on a truss framed leg. The rig worked very successfully for several years but was lost during a move in the Gulf of Mexico. The Scorpion was closely followed by The Offshore Company Rig No. 54. For Rig No. 54, however, a hydraulic jacking system on a trussed leg was used. These jack-ups were followed by Mr. Gus II, a mat supported unit using a hydraulic jacking system, which was built by Bethlehem Steel Corporation.

The early breed of jack-ups was primarily designed to operate in the U.S. Gulf of Mexico area in water depths up to 200 feet. Wave heights in the range of 20 to 30 feet with winds up to 75 mph were considered as design criteria for these units. In most cases, in the event of a pending hurricane, the rigs were withdrawn to sheltered areas. Today's jack-ups, however, are being used in international waters in a range of environmental conditions that 10 years ago were considered to be unrealistic. For example, a rig designed for 250 feet of water will have to meet the following range of criteria:

   a. U.S. Gulf Coast—55 foot wave, 125 mph wind, minimal current.
   b. North Sea—75 foot wave, 115 mph wind, 1 to 2 knot current.
   c. Southeast Asia—30 foot wave, 100 mph wind, minimal current.

As the water depth increases, the criteria rise accordingly and for 300 foot water depths the range becomes:

   a. U.S. Gulf Coast—65 foot wave, 125 mph wind, 1 to 1½ knot current.
   b. North Sea—90 foot wave, 125 mph wind, 2 to 2½ knot current.
   c. Southeast Asia—50 foot wave, 115 mph wind, ½ to 1 knot current.

These figures, although obtained from reliable sources, should not be considered finite. Actual criteria must be deter-

Fig. 1–6   *LeTourneau's Scorpion.*

mined from weather organizations in the actual geographical
location of drilling. However, the differential in criteria can
easily be seen.

A new breed of jack-ups (Figure 1-7) has been developed that
will operate in water depths in excess of 400 fett, although in
1974 the maximum criteria unit was only operating in 350 feet of

Fig. 1–7   *ETA's Europe Class Jack-up.*

water in the U.S. Gulf Coast, and in less than 300 feet in the North Sea.

Jack-up designs can generally be classified into two basic categories (Figure 1-8): independent leg jack-ups and mat supported jack-ups. Each unit has its particular value.

Fig. 1–8   *Mat supported and independent leg jack-ups.*

The independent leg jack-up will operate anywhere currently available, but it is normally used in areas of firm soil, coral, or uneven sea bed. The independent leg unit depends on a platform (spud can) at the base of each leg for support. These spud cans are either circular, square, or polygonal, and are usually small. The largest spud can being used to date is about 56 feet wide. Spud cans are subjected to bearing pressures of around 5,000 to 6,000 pounds per square foot, although in the North Sea this can be as much as 10,000 psf. Allowable bearing pressures must be known before a jack-up can be put on location.

The mat supported jack-up is designed for areas of low soil shear value where bearing pressures must be kept low. The mat is connected to all of the legs. With such a large area in contact with the soil, bearing pressures of 500 to 600 psf usually exist.

An advantage of the mat type jack-up is that minimum penetration of the sea bed takes place, perhaps 5 or 6 feet. This compares with a penetration of perhaps 40 feet on an independent leg jack-up. As a result, the mat type unit requires less leg than the independent jack-up for the same water depth. One disadvantage of the mat type unit is the need for a fairly level sea bed. A maximum sea bed slope of 1½° is considered to be the limit. Another problem with the mat supported unit occurs in areas where there is coral or large rock formations. Since mats are designed for uniform bearing, the uneven bottom would probably cause a structural failure.

Jack-ups can be either self-propelled, propulsion assisted, or nonpropelled. The majority of jack-up rigs are nonpropelled. The self-propelled unit, although very flexible, requires a specially trained crew of seamen as well as a drilling team.

Jack-ups have been built with as many as 14 legs and as few as 3 legs. As the water depth increases and the environmental criteria become more severe, we find that to use more than 4 legs is not only expensive but impractical. The prime forces on a jack-up are generated from the waves and currents, hence, the less exposure to the waves and currents the fewer the forces being developed on the unit. From this standpoint the optimum jack-up is the monopod (Figure 1-9) or single leg unit.

Problems other than wave forces, however, must be overcome with the monopod type unit. But in areas such as the North Sea with very rough seas there is a need for the monopod jack-up. Thus, research is now being done in this area. A monopod production platform is already being built for use in the North Sea.

When evaluating which type of jack-up to use, it is necessary to consider the following:

1. Water depth and environmental criteria.
2. Type and density of sea bed.
3. Drilling depth requirement.
4. Necessity to move during hurricane season.
5. Capability to operate with minimum support.
6. How often it is necessary to move.
7. Time lost preparing to move.
8. Operational and towing limitations of the unit.

Jack-ups currently constitute about 50 percent of the worldwide drilling fleet, with the semisubmersibles and the drillships composing the remaining 50 percent.

Fig. 1–9   *ETA Mobile Monopod.*

Semisubmersible Drilling Rigs

The semisubmersible evolved from the submersible (Figure 1-10) and many of today's semisubmersibles are designed to operate either resting on the sea bed or totally afloat. One of the earlier semisubmersibles was the Blue Water I (Figure 1-11)

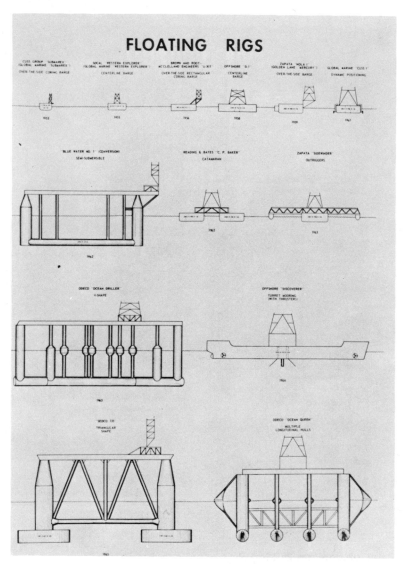

Fig. 1–10   *Semisubmersible evolotion.*

Fig. 1–11 *Blue water I.*

which was converted in 1961 from a submersible by adding vertical columns for floatation.

Today's semisubmersibles are designed for operation in water depths up to 1,000 feet and are therefore subjected to severe sea states and high winds. The general configuration of a semisubmersible consists of two longitudinal lower hulls which are used as ballast compartments to achieve the necessary drilling draft (Figure 1-12). These lower hulls are also the primary hulls while the rig is under tow. By virtue of its size and location, the semisubmersible offers low towage resistance while providing tremendous stability.

There are other designs of semisubmersibles such as the triangular design used on the Sedco series (Figure 1-13), four longitudinal hulls used on the Odeco series (Figure 1-14), and the French-designed Pentagone rig with 5 pontoons (Figure 1-15). The Pentagone unit is possibly the most successful of the multi-hull types, offering a unique symmetry and uniformity of

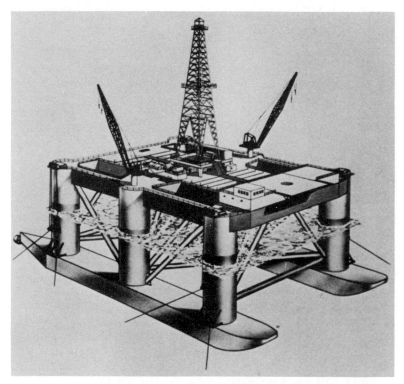

Fig. 1–12 *Bethlehem semisubmersible.*

stability characteristics. This unit does not offer the towing
capabilities of twin hull units, but it does provide good drilling
characteristics.

Semisubmersibles permit drilling to be carried out in very
deep waters and they are held on location either by a conventional mooring system or by dynamic positioning. The conventional mooring system (Figure 1-16) usually consists of 8 anchors placed in a spread pattern and connected to the hull by
chain or wire rope, sometimes even a combination of both. The
dynamic positioning method is an evolution of the ship sonar
system whereby a signal is sent out from the floating vessel to
transducers set out on the ocean floor. Dynamic positioning
becomes a greater necessity as the water depth increases and is

Fig. 1–13   *Sedco semisubmersible.*

generally considered necessary in water depths beyond 1,000 feet. However, a semisubmersible has recently been contracted for 1,500 foot water depths using the anchor and chain method. Much of the necessary chain will be carried on support vessels.

Because of the submerged mass of the semisubmersible, rolling and pitching is of a low magnitude. The motion that causes problems for the semisubmersible is heave (Figure 1-17), or the vertical motion. Because of forces on the drill string when the vessel is heaving, the semisubmersible with a low heave re-

Fig. 1–14 *Odeco semisubmersible.*

sponse is considered to be the most suitable. Heave is generated in response to exposed waterplane and is expressed as

$$T = \frac{2\pi}{\sqrt{\frac{gt}{D}}}$$

where T = time in seconds; t = tons per foot immersion; D = displacement in tons.

Therefore, the smaller the waterplane area, or 't', the lower the heave response. This is achieved in the semisubmersible by submerging the lower hulls and floating at the column or caisson level. With the loss of waterplane area to reduce heave response, a reduction in stability follows. Therefore, the de-

Fig. 1–15   *Pentagone 81.*

signer must reach a compromise between acceptable heave response and adequate stability. There are, of course, other methods of reducing heave induced forces on drill string.

Many of the new breed of semisubmersibles such as the Aker H3 units (Figure 1-18) are being designed to operate in specific areas of the world such as the North Sea where the criteria are very severe. These vessels require a very large consumables capacity, a low heave response, and good stability.

Another consideration in the design and operation of the semisubmersible is propulsion. There are several opinions on this matter, each based on valid reasoning. Propulsion is a large initial expense which can be recovered in a reasonable period of

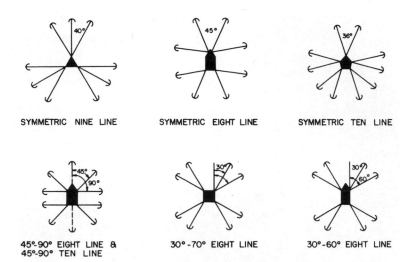

SYMMETRIC NINE LINE   SYMMETRIC EIGHT LINE   SYMMETRIC TEN LINE

45°-90° EIGHT LINE &   30°-70° EIGHT LINE   30°-60° EIGHT LINE
45°-90° TEN LINE

Fig. 1–16 *Typical mooring pattern.*

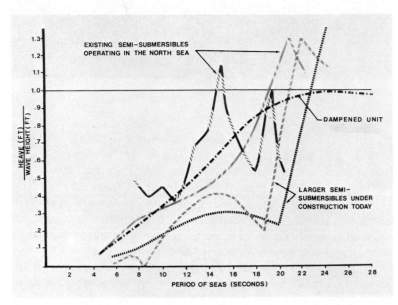

Fig. 1–17 *Heave response curve.*

Fig. 1–18   *Aker H3.*

time if mobility is required. In 1974 a large semisubmersible crossed the Atlantic in a record-breaking 21 days. The published data quote an average speed of 9.72 knots for the unit. This, of course, meant a considerable reduction in moving cost and a fair increase in productive time. On the other hand, considering that once a unit has reached location it is generally in that area for a long time, propulsion units are not only unnecessary but they also use up valuable consumables (weight) capacity.

In selecting a semisubmersible, it is therefore necessary to consider the following criteria:

a. Water depth.
b. Drilling depth requirement.

Fig. 1–20 *Glomar Challenger.*

Rig Casualties

As a point of interest, the latest information available to date
on mobile offshore drilling unit accidents still rates the jack-up
as having the most casualties, with the semisubmersible second
and the drillship third. However, when comparing the total
number of rigs built to the number of casualties by rig type
(Figures 1-23 and 1-24), the casualty rates of the jack-up and
semisubmersible are not as far apart.

For example, there have been 47 jack-up casualties from 1955
to 1974 amounting to approximately $122 million in damage
and the total number of jack-ups was 143 (Table 1-1). For semisubmersibles, there were 12 casualties amounting to about $50
million in damage out of a total of 72 semisubmersible rigs.
These figures indicate that one out of every 3 jack-up rigs ex-

Fig. 1–21   *Offshore Discoverer.*

perienced a casualty, whereas one out of every 5 semisubmersibles were involved.

When dividing the number of casualties into the total costs of damages (Figure 1-25) the average loss per rig for jack-ups was $2.76 million and the average cost for semisubmersibles was $5.66 million. In conclusion, the jack-up drilling rig may have a higher casualty rate but there are almost twice as many jack-ups as semisubmersibles and the cost of damages per rig are less for the jack-up than for the semisubmersible. Jack-up rigs also rep-

Fig. 1–22 *Offshore's Super Discoverer.*

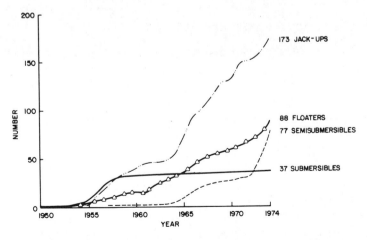

Fig. 1–23 *Cumulative number of rigs built by year and by type.*

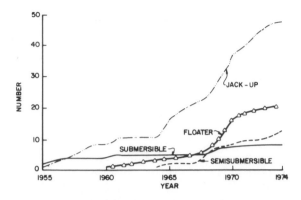

Fig. 1–24   *Cumulative number of rig casualties (by rig type).*

resent about 50% of the total rig fleet, 58% of the total value of rig casualty damages, and only 60% of the total rig casualties. All three figures show the close relationship of the number and cost of casualties to the total number of rigs.

Overall drilling rig casualties, however, have decreased. In 1957, for example, casualties occurred to about 7% of the total rig fleet, but in 1973 only 1.47% of the fleet was involved in a mishap, and it dropped to about 1.2% in 1974 (Figure 1-26).

Rig Construction

The future of the offshore mobile drilling unit industry is extremely bright. The forecast for construction shows a very

TABLE 1-1

Rig Casualties

	Total Number (1955-1974)	Total Number Built (Through 1974)	Estimated Total Value of Damages (1955-1974)
Jack-up	47	143	$122 million
Semisubmersible	12	72	$ 50 million

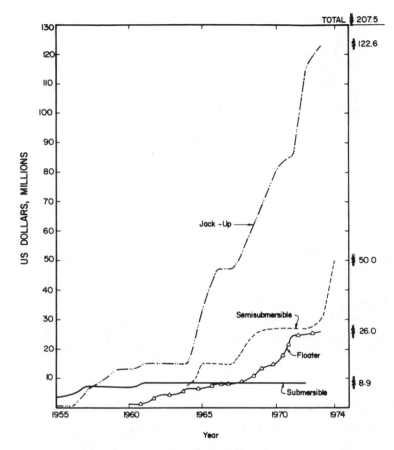

Fig. 1–25 *Estimated value of rig casualty damage (by rig type).*

high demand through 1985 for all classes of drilling units. The jack-up still leads the fleet in the total number of rigs in operation, with the semisubmersible running a distant second and the drillship far off in the rear.

In 1974 the total worldwide offshore drilling rig fleet in operation numbered 317. This includes 143 jack-ups, 72 semisubmersibles, 25 submersibles, and 77 floaters (Figure 1-27). In addition to these figures in 1974 there were 42 jack-ups under contract for

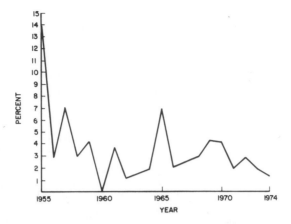

Fig. 1–26   *Casualty ratio of accidents compared to number of rigs operating.*

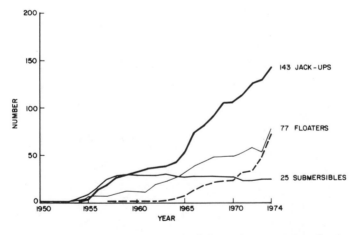

Fig. 1–27   *Cumulative number of offshore rigs operating (by type).*

construction, 75 semisubmersibles, and 18 floaters. Forty-five of the semisubmersibles are destined for use in the North Sea.

Rig count projections estimate that there will be 408 offshore mobile drilling units by 1985. Of these, 183 will be jack-ups, 121 will be semisubmersibles, and 104 will be floaters (Figure 1-28).

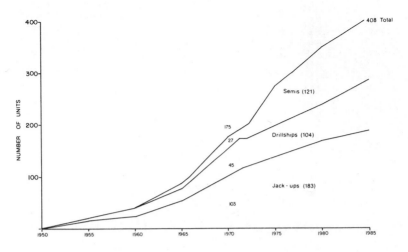

Fig. 1–28   *Mobile offshore drilling equipment.*

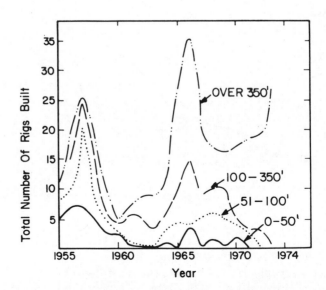

Fig. 1–29   *Annual construction of rigs by water depth capability.*

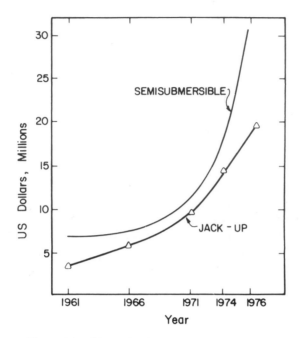

Fig. 1–30 *Chronological rig cost comparison.*

Rig construction costs continue to increase as rig designs are "scaled up" for operation in deeper water depths (Figure 1-29) under more severe criteria. Increasing numbers of rigs are being built for over 350 foot water depths. The larger and heavier rigs require increased amounts of more expensive steel and costs for labor and materials continue to spiral. For example, construction costs in 1974 were approximately six times the costs in 1961. The average cost of construction for a jack-up rig increased from $3.3 million in 1961 to $14.5 million in 1974 and the cost is expected to be approximately $19.65 million in 1976 (Table 1-2 and Figure 1-30).

For the semisubmersible, the average construction cost in 1961 was $6.67 million. The cost in 1974 was about $20.5 million and it is expected to be around $30.76 million in 1976. These construction cost escalations are not unreasonable. In fact they may prove to be conservative. Much of this cost is due to the size and sophistication of the newer drilling rigs.

TABLE 1-2

Average Construction Costs (U.S. $ Million)

	1961	1966	1971	1974	1976
Jack-up	3.3	5.15	9.75	14.5	19.65
Semisubmersible	6.67	7.57	13.2	20.5	30.76

Because of increased construction costs, sophistication of equipment, and specialized operating personnel the day rates for the new rigs are also skyrocketing. Day rates in the past were based on investment and depreciation of equipment. Now the market is vulnerable and day rates are being decided more by demand than by dollar investment. The market has changed and the industry must change with it.

2

Stability and Moving

Ralph G. McTaggart
ETA Engineers, Inc.

Stability

Stability is simply the ability of a rig to remain afloat. The subject of stability is further divided under the two headings of (a) Intact Stability and (b) Damaged Stability.

For every rig, the designer or builder should furnish the rig owner with a Stability Booklet which, at a very minimum, should contain (a) Hydrostatic Properties), (b) Cross Curves of Stability, (c) Statical Stability Curves, and (d) Dynamic Stability Curves. Items (c) and (d) should be sufficient to cover the normal operating range of the vessel.

A brief explanation of the above items follows:

a. *Hydrostatic Properties* (Figure 1-31) are generated from the shape of the underwater portion of the rig and can be used to determine the weight of the rig and the location of the centroid, longitudinally and transversely. It has many other uses, but the ones just mentioned will be most often used when moving a rig.

b. *Cross Curves of Stability* (Figure 1-32) are also generated from the underwater portion of the rig and are used by the designer to determine the amount of stability the vessel has when it is not in the upright position. Figure 1-33 shows a typical transverse section through a hull. This figure illustrates how the value of GZ is determined. As the KG value increases the GZ value decreases, or vice versa.

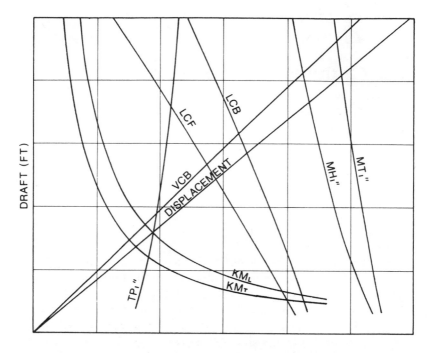

Fig. 1–31 *Hydrostatic properties.*

c. *Statical Stability Curves* (Figure 1-34) are developed from the cross curves of stability and are curves of righting arm. They are sometimes referred to as GZ curves.

d. *Dynamic Stability Curves* (Figure 1-35) are produced from the statical stability curves and calculations to determine the overturning moment caused by a wind of a given velocity. This curve is probably the most significant of all curves because it shows whether or not the rig can be towed during the forecasted weather, while remaining within the safety parameters of the regulatory bodies.

Other information that many would find useful and could be included in the Stability Booklet, but which is not considered to be essential, includes (e) Allowable Dynamic Stability KG Curves, (f) Damaged Stability Calculations, (g) Motion Response Analysis, and (h) Lightship Characteristics.

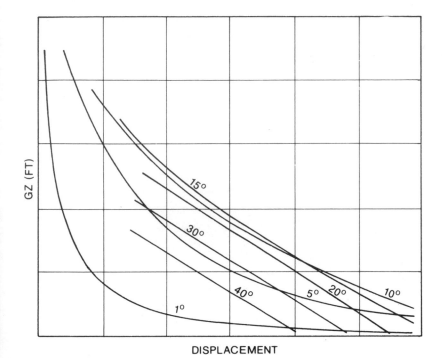

GZ (FT)

DISPLACEMENT

Fig. 1–32 *Cross curves of stability.*

e. *Allowable Dynamic Stability KG Curves* (Figure 1-36) are generated from the dynamic stability calculations. These curves are a growth of the dynamic stability curves, and they simplify the rig mover's job by eliminating the need to prepare a calculation of dynamic stability every time he decides on a possible tow condition.

f. *Damaged Stability Calculations* should be prepared for the effect of damage to the outside compartments or flooding into any compartment. These calculations should show that the vessel has sufficient reserve stability to survive either damage or flooding. If the ABS "Rules for Building and Classing Offshore Mobile Drilling Units 1973" are applied to the vessel, the ability to survive damage or flooding must be considered in association with the overturning effect of a 50-knot wind.

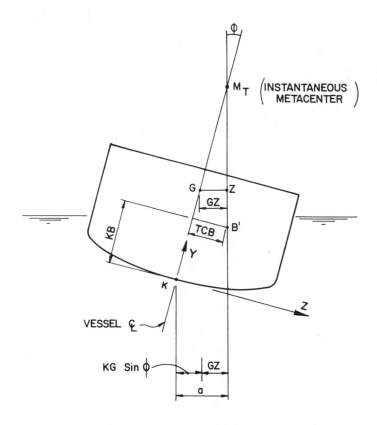

Fig. 1–33 *Transverse righting arm and center of buoyancy.*

g. *Motion Response Analysis* is the study of the rig in a "hove
   to" state. This is the position when "going on location,"
   and the results of this analysis are used to determine the
   stresses induced when a jack-up leg touches bottom or
   those caused by mooring forces on a drillship or semisubmersible.
h. *Lightship Characteristics* are probably the most used (or
   misused) information that should be supplied. This information is prepared either from a series of accurate weight
   calculations or from an inclining experiment, or both. The

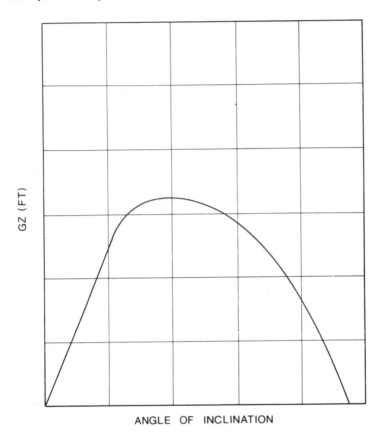

ANGLE OF INCLINATION

Fig. 1–34 *Statical stability curve.*

calculations determine the weight and center of gravity in all directions of the dry rig, i.e., no variables of any kind are included. From this information, the operator determines the condition of the rig at any time. It must be stressed here that although the shipbuilder may have gone to great lengths to determine the lightship upon completion, it is up to the owner or operator to make sure that the values are adjusted if any changes are made to the rig, i.e., equipment or structure changed, added, removed, or even relocated. The Lightship Characteristics are the foundation of all

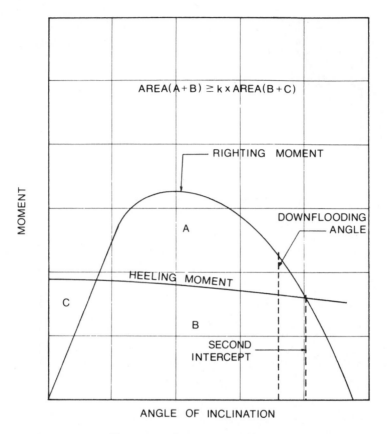

Fig. 1–35 *Dynamic stability curve.*

calculations for the afloat and elevated positions, and an inaccurate number not only makes all other calculations worthless, but could also endanger the safety of the rig and personnel.

Moving

Moving is the intentional relocation of the rig for any purpose, though we think of changing drilling location when we talk of

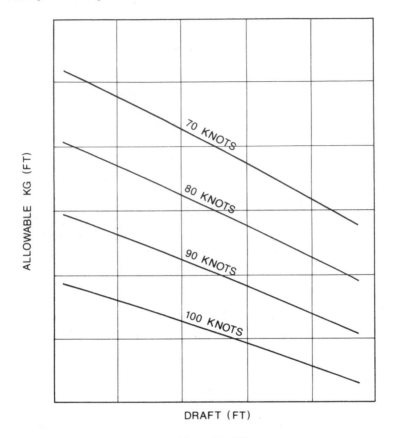

Fig. 1–36 *Allowable KG curve.*

moving. The prime consideration in preparing to move and while moving must be safety. Therefore, it is essential that the rig mover be completely familiar with the rig and the expected area environment during the move. He should have a basic knowledge of Naval Architecture and know how to apply it to his vessel.

It has not been the practice in the past for the rig mover to discuss the rig with the designer, or builder's naval architect, prior to delivery of the rig. I feel that a discussion between these people would not only increase the education of both, but would also help ensure a safer operation and, I would hope, eliminate disasters due to incompetence.

Every rig owner should receive from the builder a book entitled "Operating Book" or "Booklet of Operating Conditions." The rig mover should read and understand this book before attempting to move the rig. Each rig is like a new car, and even though you know how to drive, each car has its own peculiarities—so also does each rig, even so-called "sister ships."

Again let me stress to the rig mover: understand your rig, talk with the Naval Architect, and read the "Operating Book."

Moving comes under two basic categories: (a) Field Transit and (b) Ocean Tow. A Field Transit condition is generally considered to be a move that would require no more than a 12-hour voyage to a location where the unit could be jacked up, or to a protected location. An Ocean Tow should be considered for all other moves.

Preparation for moving a semisubmersible rig or drillship is covered by the "Operating Book," but basically it would be as follows:

1. Give notices to parties concerned.
   A. Provide overall report for Owner.
   B. Give insurance company and their representatives the following:
       (a) Distance of move, miles
       (b) Area and water depth of next location
       (c) Number, size and ownership of tugs to be used
       (d) Name of person in charge of move
   C. Provide for towing company the following:
       (a) Distance of move, with initial and final positions
       (b) Number, horsepower, and type of tugs
       (c) Required towing equipment
       (d) Radiofrequency for communications
       (e) Name of person in charge of move
       (f) Request for tugs to contact rig, giving estimated time of arrival, names of captains and tugs, and which tug is lead

2. Obtain long-range weather forecast.
3. Finalize move schedule.
   A. Complete stability calculations for upcoming move.
   B. Develop deballasting schedule.
4. Dispatch tugs and anchor-handling crew.
5. Tie down all cargo.
6. Man pump rooms.
7. Deballast, as scheduled, to desired draft and recheck calculations.
8. Secure pump rooms.
9. Connect tugs.
10. Disconnect anchors, commence tow, give U.S. Coast Guard pertinent data regarding tow and notify USCG when completed.
11. Move into anchor pattern, connect anchors, disconnect and release tugs not being used for resetting anchors.
12. Finalize ballast schedule to drilling draft, recalculate stability.
13. Man pump rooms.
14. Ballast to drilling draft (close all valves, take soundings of all tanks).
15. Complete stability calculations at drilling draft.
16. Pretension anchors and release to operating tensions for 1-hour static test, all activity ceased.
17. Release remainder of tugs.
18. Retake soundings of all tanks to make sure there are no leaks.
19. Secure pump rooms.
20. Authorize drilling operations to commence.

The following is a sample preparation list for floating, raising, lowering, and preloading a jack-up:

1. Check operator's console, main switchboard, and accessory equipment for corrosion, loose wiring, damage, etc. that may result in malfunction. Check all other electrical and mechanical components.
2. Put all generators on line to check paralleling characteristics and overall engine-generator operations.

3. Check all electrical circuits for faulty insulation, breaks, and grounds.

4. Check all motor and brake functions and wiring if they have not been satisfactorily operated within the past 24 hours. Check and record all elevating motor torques.

5. After checking all brakes, motors and torque readings, reinspect each motor to assure that all of the brake plugs have been reassembled and are tight. Replace motor covers.

6. During all operations, the platform should be kept trim. Make level or equal-draft reading when approaching or leaving drilling location and check for an equal loading on each spud can when elevated and when elevating or lowering the platform. Keep a log of these conditions and record at least every 12 hours so necessary steps can be taken to maintain a trim condition.

7. Check anchors, fairleads, anchor lines, and winches for proper operation. Check motor power and free-spooling.

8. All service tanks should be either "pressed-up," full, or emptied of liquids. This may not always be practical for potable water and diesel-fuel systems; however, do not exempt more than one tank in each of these systems from the rule.

9. All below waterline discharges should be checked and closed.

10. Check diesel fuel tanks and drain off water. Check water-fuel filters and air filters.

11. Check bilge system and all liquid-service systems. Inspect for "frozen" valve and safety devices. Voids should be clean.

12. Periodically check derrick-holding devices.

13. Before lowering platform, be sure jet hoses (one for each jet line) are readily available and in operating condition. Inspect deck manifolding and water-pumping arrangements to make sure they are operable.

14. Keep emergency and repair supplies aboard at all times. Periodically inspect "stores" and note where they are kept for fast, efficient use. Rolls of heavy plastic, rope, and tape

are good emergency repair items for hatches, etc. Welding, burning, and cutting equipment should be kept readily available.

15. Tie down derrick block, swivel, and all movable items on the derrick floor.

16. Tie down drill pipe, collars, casing, and all movable items on the deck. Check for loose boxes, crates, wires, gas bottles, etc., both on deck and within the hull. Secure cranes and booms.

17. Close and secure hatches, doors, manholes, return lines, and all other openings between the outer and inner hull that are not in use. Station personnel near openings that are required to stay open during a move so that they may be immediately closed in an emergency. Keep the platform in this condition during all moving and preloading operations.

18. Prepare the "sea water" tower for operation when the platform is raised or lowered.

19. Check adequacy of tugs and towing equipment.

20. Check and record weather data. Use more than one source when available. Do not move when good weather and seas are estimated to be of short duration.

21. Periodically inspect all safety, aids-to-navigation, and lifesaving equipment.

23. Check ocean bottom data at the proposed location to determine bottom conditions.

24. Prior to lowering spuds, check wind-direction, current-direction, and depth-recorder readings.

25. Record spud penetration data and use as references when removing spuds from ocean floor.

26. Know your personnel and your platform.

27. Confine raising or lowering of platform to periods of good visibility, preferably during daylight.

The above list is an attempt to cover a general area of moving checks. It is the responsibility of the rig mover to ensure a safe operation and to check the list given in the "Operating Book."

Hazards

The hazards that might be expected during a tow are too numerous to list, but I shall cover a few. Probably the most significant hazard occurs when the rig is preparing to "go on" or "come off" location. In the case of jack-ups, consideration must be given to the sea state because of the change in floating characteristics that takes place while moving the leg up or down.

The effect of the leg striking the ocean floor must also be considered. This should have been taken care of by the designer, and the "Operating Book" should have a section related solely to allowable conditions for going on and coming off loction. For the drillship and the semisubmersible the sea state must also be considered, but for a different reason. Anchor handling and the effect of unequal mooring arrangement should be considered. A mooring arrangement and procedure should be included in the "Operating Book."

Also included in the "Operating Book" should be the limits of service in both the operating and towing positions. It is false and senseless to ignore this section because to do so can only endanger the rig and personnel. If it is necessary to stray from the design criteria, a few dollars spent in engineering analysis shall serve to ease many minds and to reduce the risk potential.

One of the greatest fears that develops while at sea is that of damage which may produce flooding and, if extensive enough, the loss of the rig. It is easy to say that with a little common sense most damages and flooding can be avoided, but they still occur. The designer knows this, and should have compartmented the rig such that flooding shall be contained within an allowable extent, which may be uncomfortable to those on board, but shall not result in a loss of the rig. On most drilling rigs presently in service, excluding drillships and self-propelled units, the probability of damage due to collision is remote. And if it takes place, it will be of such low impact velocity as to cause only a slow leak rather than a rapid flood.

The effect of the vertical center of gravity on a damaged vessel is considerable. If damage should occur, steps should be taken to lower the vertical center of gravity. This can easily be done on a jack-up by lowering the legs. On the other vessels ballasting can improve the stability. However, this is a condition that can occur suddenly, and it is not always possible to carry out a damaged stability calculation when the water is entering your cabin. Here again we see the need for preplanning, and this should be discussed between the Owner and the Naval Architect.

Problems do, of course, occur without damage. A sudden squall, a change of sea state, or a wind coming out of nowhere can cause many things to happen. Even the mightiest rig can act like a cork in the ocean on such occasions. Fortunately, we have more accurate weather forecasting today, and the rig mover sometimes can prepare for the change. Several courses of action are available: a jack-up can lower the legs, a semisubmersible can ballast, and a drillship or self-propelled vessel can alter course to avoid, or at least reduce, the effect of the squall or other problems. Once again, each vessel has its own peculiarities, and consultation with the designer combined with the experience of the rig mover can usually serve to eliminate serious conditions.

If a search were made for the major contributing factor in accidents occurring during tow, it would be found that bad design and inexperience head the list. It should be remembered that the offshore drilling rigs are still new in comparison to ships. The designer can only learn from feedback, but fortunately, technology is advancing rapidly in the marine field and information is more readily available today.

In conclusion, I would again stress the importance of discussions between the designer and operator on the peculiarities of each rig. Education can only improve safety. Safety can only mean more operating time and more return on dollar investment.

Nomenclature

Draft:	Depth of submerged hull
LCB:	Longitudinal center of buoyancy
VCB:	Vertical center of buoyancy
LCF:	Longitudinal center of floatation
Displacement:	Weight
KM_L :	Longitudinal metacentric height above keel
K:	Keel

KG: Vertical distance from keel to center of
                       buoyancy
M: Metacenter
KM_T: Transverse metacentric height above keel
MH1″:   Moment to heel one inch
MT1″:   Moment to trim one inch
TP1″:   Tons per inch immersion
GZ: Heel lever arm
G: Location of center of gravity
Righting Moment: Displacement multiplied by GZ
Heeling Moment: Overturning moment produced by wind
Downflooding Angle: Angle of heel at which water will enter the hull
                       through an opening
Second Intercept: Second crossing of righting moment and heel-
                       ing moment curves
k: Constant, as defined by regulatory agency.

3

Wind, Wave, and Current Forces on Offshore Structures

Odd A. Olsen
Det norske Veritas

The forces produced by wind, wave, and current are the primary design loads on mobile drilling units and other offshore structures. These forces are dynamic and ever-changing; rarely can they be expressed as a mathematical function of time. They are statistical in nature and should, if possible, be handled by means of statistical tools.

The most commonly used method so far for evaluating wave loads on an offshore structure has been to base the calculations on one or more design waves of specified height and period. This was also the practice in the shipbuilding industry some years ago. When ship sizes increased, however, it became obvious that the design wave method led to unreasonable results in many cases. It was realized that the random nature of the ocean waves and consequently the responses of a structure to these waves could only be described by statistical methods.

Research carried out several years ago revealed that the irregular wave pattern could be described linearly by superimposing regular waves of different frequencies, heights, and directions of travel. In this way, it was possible to take into account all the wave components present in a realistic sea. The development of methods for calculating the response of a floating structure to regular waves was also initiated in this period. Det norske Veritas (DnV) was involved in this development and produced a number of computer programs related to ships' problems.[1,2]

Calculations based on the principles indicated proved successful for ships and these methods are now commonly used. It was a natural step to extend the same principles and methods to offshore structures, such as mobile drilling rigs.[3]

The wave loads on both floating and fixed offshore structures may in principle be handled by use of the methods described above. Considering fixed offshore structures, such as jacket type structures or gravitational structures of the caisson type, this method requires that the nonlinear loads (for instance, drag forces) be small in comparison to the linear loads (for instance, inertia forces). If not, nonlinear statistics have to be considered, making it more complicated to arrive at a useful solution.

Yet no such method for practical application exists that takes into account the nonlinearity of the loads. Consequently, when these nonlinearities are of significant influence to the overall loads, the old design wave method has to be applied in the analysis and design procedure.

Thus, the structure is designed for a specified wind and current velocity as well as a wave of specified height and period. These criteria describe the "50 year storm" or the "100 year storm." This refers to the worst wind, wave, and current conditions expected in any arbitrarily chosen 50 or 100 year period. At DnV, the "100 year storm" has been chosen as the design criterion.

Although the design wind and wave conditions are not uniform, they reflect the expected maximum values based on long term statistics. Typical design criteria (storm condition) for fixed offshore structures intended to operate in two different areas of the world are given below:

	Gulf of Mexico	North Sea
Water depth to L.A.T.	84m.	92 m.
Wave height	19 m.	30 m.
Wave period	15 sec.	16 sec.
Wind velocity	50 m./sec.	60 m./sec.
Current velocity at surface	0-0.1 m./sec.	1.4 m./sec.

In summation, the two methods normally used in estimating the wave loads on fixed and floating offshore structures and some of their characteristic features are:

Spectral Analysis Method

- Used for floating structures mainly, but also fixed structures provided that the nonlinear loads are small compared to the linear loads (large diamter columns).
- Linear statistic analysis.
- Evaluation of the most probable largest wave loads that, on the average, will occur during the structure's lifetime.

Design Wave Method

- Used for both fixed and floating offshore structures.
- Design wave of specified height and period.
- Evaluation of the loads resulting from a regular wave with height and period as specified.

Applying the design wave method, the combined effects of wind, waves, and current on a fixed offshore structure are normally treated as quasistatic forces. This is valid in most cases, particularly if care is taken in determining the "worst case" loading. The environmental loads acting on different drilling units are simplified in Figure 1-37.

WIND

Wind velocity is an important parameter in wind forces. Normally, two different wind velocities are specified in the design criteria for offshore structures, the N year sustained and the N year gust wind velocity. These are defined as follows:

N Year Sustained Wind Velocity

This is the average wind velocity during a time interval (sampling time) of one (1) minute with a recurrence period of N years.

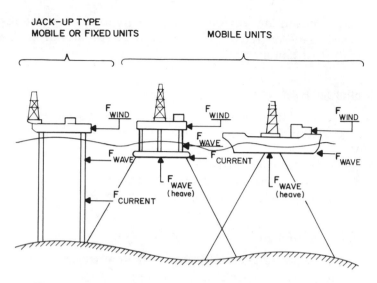

Fig. 1–37   *Environmental forces acting on drilling units.*

N Year Gust Wind Velocity

This is the average wind velocity during a time interval (sampling time) of three (3) seconds with a recurrence period of N years.

At DnV, the 100 year wind velocity is used as the design criterion. This velocity may be evaluated from wind statistics for the specified area. Which wind velocity, however, gust or sustained, is to be used in force calculations? DnV has adopted the following principle:

> If gust wind alone is more unfavorable than sustained wind in conjunction with wave forces, the gust wind speed is to be used. If the sustained wind and wave forces are greater than gust wind alone, the sustained wind speed should be used.

It should be mentioned that recent research by the author has shown that the long term distribution of wind speed may be described by a Weibull distribution function (defined in the section on "Waves").[4]

Velocity Variation with Height above Sea Surface

The sustained wind velocity is often considered to vary with height above sea surface according to the following formula[5]:

$$V_z = V_{10} \left(\frac{Z}{10}\right)^{1/\tau} \tag{1}$$

Davenport[5] has proposed a value of $\tau = 7$ for open country, flat coastal belts, and small islands situated in large areas of water. Davenport does, however, further state that if there are areas in which the highest probable velocities occur during severe local storms such as thunderstorms and frontal squalls, little or no increase in velocity with height would seem appropriate.

The 100 year storm in the North Sea must be considered to be a severe local storm, and, according to Davenport, little or no increase in wind velocity with height should be expected, especially when gust velocity is considered.

The following formula, adopted as basis for the DnV Rules[6,7], is a compromise between the one-seventh power law (Equation 1) and Davenport's conclusion:

$$V_z = V_{10}\sqrt{0.93 + 0.007\,Z} \tag{2}$$

(It should be mentioned that other authorities use the one-seventh power law.)

According to the DnV Rules[6,7], the gust velocity is not to be taken less than:

$$\left(V_z\right)_G = V_{10}\sqrt{1.53 + 0.003\,Z} \tag{3}$$

The gust wind velocity as determined by Equation (2) varies less with height than does the sustained wind velocity.

The three velocity profiles defined by Equations (1), (2), and (3) are shown in Figure 1-38.

Wind Forces of Individual Members

Wind forces have been found to increase with the square of the wind velocity and in direct proportion to the exposed area.

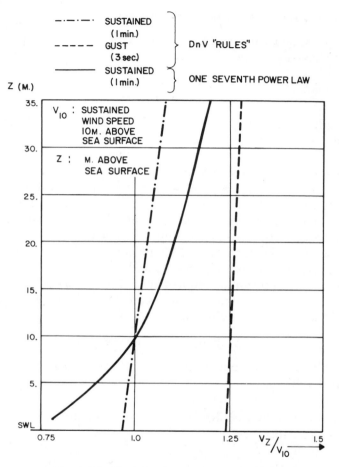

Fig. 1–38 *Variation in wind speed with height.*

Other factors which affect the wind forces are the shape of the exposed areas and their height above sea level.

Based on the one-seventh power law, the formula for calculating wind forces may be written as follows:

$$F_w = 0.00388 \cdot C_H \cdot C \cdot \overline{V}_{10}^2 \cdot A \qquad (4)$$

(The units are ft., m., lbs., and sec.)

The height coefficient C_H may be expressed as:

$$C_H = \left(\frac{Z}{30}\right)^{2/7} \qquad (\text{where } Z \text{ is in ft.}) \qquad (5)$$

The shape coefficient C varies from 0.5 for cylindrical shapes to 1.5 for structural steel shapes.

The force acting on the area A in the direction of the wind then becomes:

$$F_y = 0.00388 \, C_H \, C \, V_{10}^2 \, A \, \text{Sin}^3 \, a_1 \qquad (6)$$

The formula generally accepted at DnV for calculating wind forces is:

$$F_w = C \, \frac{V_z^2}{16} \cdot \text{Sin} \, a_1 \cdot A \qquad (7)$$

(The units are metric.)

The force acting on an area A in the direction of the wind then becomes:

$$F_y = C \, \frac{V_z^2}{16} \, \text{Sin}^2 \, a_1 \cdot A \qquad (8)$$

As may be seen from Figure 1-39, these two formulae differ by $\sin \alpha_1$. The question of which formula gives the best estimate of reality has been long discussed. The disagreement between authorities is great. DnV has adopted the conservative estimate both because it is conservative and because they feel it is closer to reality.

Shape Coefficient

The shape coefficient for short individual members (3-dimensional flow), according to the DnV "Rules," is to be taken as:

$$C = C_\infty \left(0.5 + 0.1 \, \frac{L}{d}\right) \qquad (9)$$

This formula is to be applied for $\frac{L}{d} < 5.0$.

The shape coefficient C_∞, according to the DnV "Rules," is shown in Figure 1-40.

A) DECOMPOSING VELOCITY FIRST

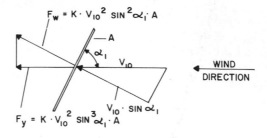

B) DnV APPROACH

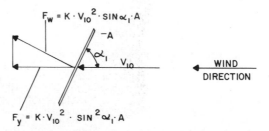

Fig. 1–39   *Two different methods for decomposing wind or drag forces.*

Total Wind Forces

The total wind force acting on a structure is determined by summing up the effects of wind forces acting on separate areas of the structure. The total wind force on a typical jack-up rig is shown in Figure 1-41. Note that the force due to a 56.3 m./sec. wind velocity is twice the force due to a 40 m./sec. wind velocity.

It should be emphasized that wind direction can be a factor in determining the "worst case" loading on a structure.

CURRENT

Current forces are often considered in connection with wave force by adding vectorially the water particle velocities due to

DNV "RULES"

MEMBER	RESTRICTION	C_∞
FLAT BARS, ROLLED SECTIONS, PLATE GIRDERS AND OTHER SHARP−EDGED SECTIONS	SHARP− EDGED SECTIONS	2.0
RECTANGULAR SECTIONS	 A/B ≥ 2.0	1.5
SMOOTH CIRCULAR CYLINDRICAL MEMBER	D ≥ 0.3 M.	0.7
	D < 0.3 M.	1.2

i) FOR DECKHOUSES AND SIMILAR STRUCTURES
    PLACED ON HORIZONTAL SURFACE THE FORMULA
    FOR C IS APPLICABLE TAKING ℓ AS THE
    VERTICAL DIMENSION AND d AS THE HORIZONTAL
    DIMENSION OF CONSIDERED AREA.

    FOR C_∞ USE COEFFICIENTS GIVEN IN THE
    TABLE ABOVE.

Fig. 1–40 *Shape coefficient (2-dimensional flow).*

wave and current. Therefore, only the current velocity profile to be applied is considered in this section.

Current velocity is often separated into two parts, wind-induced current due to wind shear and tide-induced current.

The variation of current velocity with distance above bottom may, if detailed field measurements are not available, be described by the following equation:

$$V_{cy} = V_w\left(\frac{y}{H_1}\right) + V_T\left(\frac{y}{H_1}\right)^{1/7} \tag{10}$$

In open areas, if statistical data are not available, the wind-induced current of the still water level may be taken as:

$$V_w = 0.01\, V_{10} \tag{11}$$

The tide-induced current should be chosen according to field measurements, if available.

WIND FORCE kp·10⁵

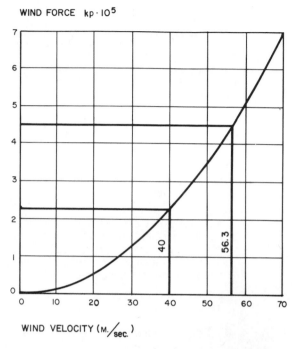

WIND VELOCITY (M./sec.)

Fig. 1–41   *Wind flow on jack-up drilling platform.*

The velocity profiles given above are shown graphically in Figure 1-42.

Recent research indicates that it may not be correct to add the current velocity vectorially to the water particle velocity in the wave.[8] The results seem to be too conservative when the current is acting in the direction of the wave propagation.

When the current is acting in opposite direction of the wave propagation (tidal current only), the interaction between current velocity and water particle velocity may lead to more severe conditions than that obtained by vectorially adding the velocity components.

It should be emphasized that this condition will occur only occasionally as the wind-induced current in the direction of the wave propagation is normally larger than the tide-induced current in the opposite direction.

$$V_{cy} = V_T \cdot \left(\frac{y}{H_I}\right)^{1/7} + V_w \cdot \left(\frac{y}{H_I}\right)$$

V_{cy} : CURRENT VELOCITY y
M. ABOVE BOTTOM (M./SEC.)

V_T : TIDE- INDUCED CURRENT
IN THE SEA SURFACE (M./SEC.)

V_w : WIND—INDUCED VELOCITY
IN THE SEA SURFACE (M/SEC.)

H_I : WATER DEPTH (M.)

y : M. ABOVE BOTTOM

V_{IO} : SUSTAINED WIND SPEED (M/SEC.)

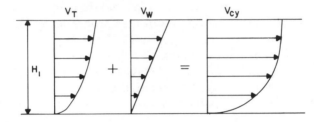

Fig. 1–42 *Current velocity due to wind and tide.*

WAVES

As mentioned earlier, two different methods are normally used in evaluating the wave forces on fixed and floating offshore structures.

Spectral Analysis Method

The spectral analysis method was earlier said to be not very useful for fixed offshore structures when the nonlinear loads (drag forces) are important. When considering floating offshore structures, however, the spectral analysis method is commonly used. Its basic principles are outlined below.

1. *Regular Waves*

Regular waves do not exist in reality. They are interesting, however, as an irregular wave system may be simulated by a large number of regular waves. A regular deep water wave is approximately sinusoidal and the following relationship exists between wave length and wave period (metric units):

$$\lambda = \frac{g}{2\pi} T^2 \cong 1.56 \, T^2 \qquad (12)$$

It may also be shown that the maximum height of a regular wave is given by:

$$H_{max.} = \frac{1}{7} \, \lambda \qquad (13)$$

The wave will break before its height increases.

The responses of floating offshore structures to regular waves of different frequencies, heights, and directions can be calculated. Several computer programs for this calculation have been developed at DnV [9,10].

As the responses to regular waves are approximated very well by linear functions of wave height, these responses may be represented by the so-called transfer functions. Typical examples of transfer functions are shown in Figures 1-43 through 1-48.

A transfer function gives us information about the response to regular waves only. Phase information is also usually available together with the transfer function (not incorporated in Figures 1-43 through 1-48.

2. *Irregular Waves*

To obtain a realistic description of the sea, irregular waves have to be considered. These are more difficult to describe than the regular waves because the wave pattern never repeats itself. Statistical methods must therefore be used to describe the sea. It should be emphasized, however, that these methods are not less exact. The waves are statistic in nature and should, if possible, be described by use of statistical tools.

In describing irregular waves, one distinguishes between stationary conditions during a short time ($\approx$ hours) and variable conditions during a long time ($\approx$ months or years).

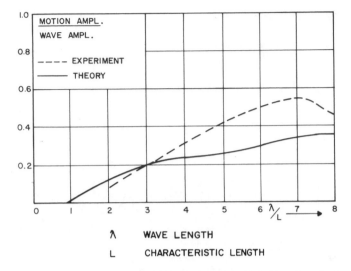

Fig. 1–43 *Surge transfer functions, Aker H-3, Heading β=45°*

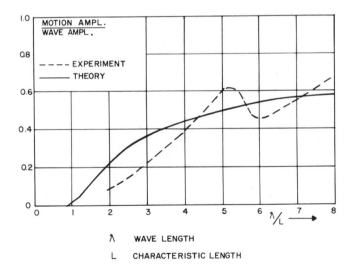

Fig. 1–44 *Sway transfer functions, Aker H-3, Heading β=45°*

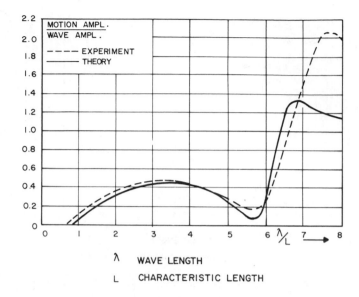

Fig. 1–45 *Heave transfer functions, Aker H-3, Heading $\beta=45°$*

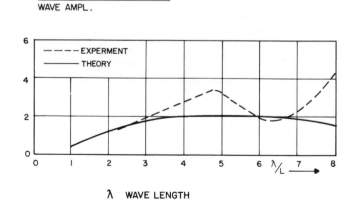

Fig. 1–46 *Roll transfer functions, Aker H-3, Heading $\beta=45°$*

ANGLE AMPL.X WAVE LENGTH
WAVE AMPL.

λ   WAVE LENGTH
L   CHARACTERISTIC LENGTH

Fig. 1–47   *Pitch transfer functions, Aker H-3, Heading β=45°*

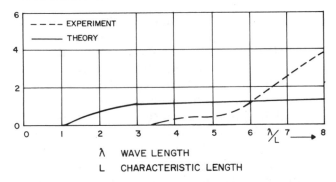

ANGLE AMPLITUDE X WAVE LENGTH
WAVE AMPLITUDE

λ   WAVE LENGTH
L   CHARACTERISTIC LENGTH

Fig. 1–48   *Yaw transfer functions, Aker H-3, Heading β=45°*

a) *Stationary Conditions: Short term Statistics*

Stationary conditions do not imply that the ocean surface is constant or repeats itself, but rather that the statistical parameters are constant.

An irregular sea such as that shown in Figure 1-49 may be described by linearly superimposing a great number of regular waves with different frequencies,

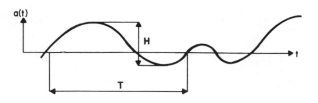

DEFINITIONS

H :	WAVE HEIGHT	T :	ZERO UPCROSSING PERIOD
$H_{1/3}$:	SIGNIFICANT WAVE HEIGHT = AVERAGE OF THE ONE THIRD HIGHEST WAVE HEIGHTS	$\overline{T}$:	AVERAGE APPARENT ZERO UPCROSSING PERIOD

SHORT TERM : $\overline{T}$ AND $H_{1/3}$ ARE CONSTANTS

LONG TERM : $\overline{T}$ AND $H_{1/3}$ ARE VARIABLES

Fig. 1–49   *Irregular sea.*

phase angles, and directions. The procedure used to describe this principle is shown in Figure 1-50.

This superimposing of waves is usually done in two steps. First waves with the same direction but of different frequencies, heights, and phases are superimposed. The result becomes an irregular wave system with infinitely long crests (long-crested sea). The next step is to sum up several such long-crested wave systems with different directions (Figure 1-51). A specified percentage of energy should be given to each direction. The final result becomes an irregular short-crested wave system.

Mathematically this wave system may be described by what is known as a wave spectrum and is a product of two functions:

$$\left[S_w \left(\omega, a \right) \right]^2 = \left[S_w \left(\omega \right) \right]^2 \cdot f \left(a \right) \tag{14}$$

The wave spectrum provides information about the energy distribution in the irregular wave system. At

FREQUENTLY DOMAIN

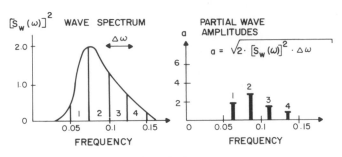

TIME DOMAIN

PARTIAL WAVES IN TIME

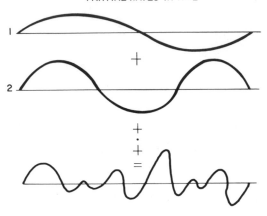

Fig. 1–50   *Simulation of waves.*

DnV, the modified Pierson-Moskowitz wave spectrum is normally used. It is defined by the following equation as a function of the circular frequency:

$$\frac{\left[S_w\left(\omega\right)\right]^2}{\left(H_{1/3}\right)^2 \cdot \overline{T}} = \frac{1}{8\pi^2}\left(\frac{\omega\overline{T}}{2\pi}\right)^{-5}\exp.\left[-\frac{1}{\pi}\left(\frac{\omega\overline{T}}{2\pi}\right)^{-4}\right] \quad (15)$$

This nondimensional spectrum is shown graphically in Figure 1-52.

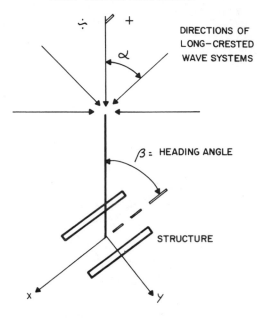

Fig. 1–51 *Wave directions.*

The directionality function provides information on the dispersion of energy in the different long-crested wave systems used in simulating the short-crested wave system. The function normally used at DnV, shown in Figure 1-53, is defined by the following equation:

$$f(a) = \frac{2}{\pi} \cos^2 a \qquad (16)$$

A basic assumption must be made in order to proceed using statistical tools.

The response amplitudes to a wave system which may be described as a sum of a great number of regular harmonic components are the sum of the responses for each component. Mathematically this may be expressed as:

$$\delta_{li} = Y_{li} \cdot a_{li} \cos(\omega_i t + \xi_{li}) \qquad (17)$$

$$\frac{\left[S_w(\omega)\right]^2}{H_{1/3}{}^2 \cdot \bar{T}} \qquad \frac{\left[S_w(\omega)\right]^2}{H_{1/3}{}^2 \cdot \bar{T}} = \frac{1}{8\pi^2}\left(\frac{\omega \cdot \bar{T}}{2\pi}\right)^{-5} \exp\left(-\frac{1}{\pi} \cdot \left(\frac{\omega \bar{T}}{2\pi}\right)^{-4}\right)$$

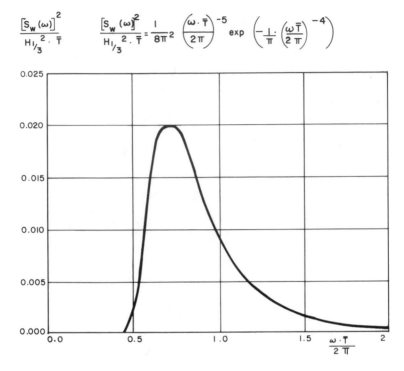

Fig. 1–52 *Modified Pierson-Moskowitz spectrum in nondimensional form.*

Analogous to a wave spectrum, a response spectrum may be defined as follows:

$$\left[S_R(\omega)\right]^2 d\omega_i = \sum_{d\omega_i} \frac{1}{2}\left(Y_i\, a_i\right)^2 \qquad (18)$$

When Y is constant within the interval $\Delta\omega_i$, we can write the following equation:

$$\left[S_R(\omega)\right]^2 = Y^2\left[S_B(\omega)\right]^2 \qquad (19)$$

This linear superimposing principle is shown in Figure 1-54.

The parameter describing the statistical distribution of the response amplitudes is directly related to the area beneath the response spectrum curve. As the wave

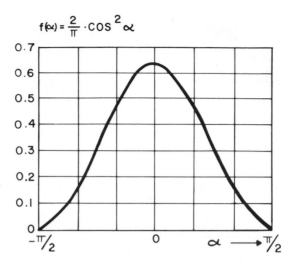

$$f(\alpha) = \frac{2}{\pi} \cdot \cos^2 \alpha$$

Fig. 1–53 *Directionality function.*

spectrum is described by the two parameters H⅓ and $\overline{T}$ related to observed wave data, the response amplitudes are calculated as a function of these two parameters.

The short term statistical distribution of response amplitudes or wave heights is assumed to be adequately described by the Rayleigh distribution function shown in Figure 1-55. This distribution is defined by the following equation:

$$P_S(\sigma) = 1 - \exp\left(-\frac{\sigma^2}{E}\right) \qquad (20)$$

The parameter, E, is related to the response spectrum by the following equation:

$$E = 2\,S_R^2 \qquad (21)$$

The most probable largest response amplitude during a time interval, t, may be expressed as:

$$\sigma_{max.} = \sqrt{E\,\ln(t/\overline{T}_R)} \qquad (22)$$

Considering wave heights the distribution function

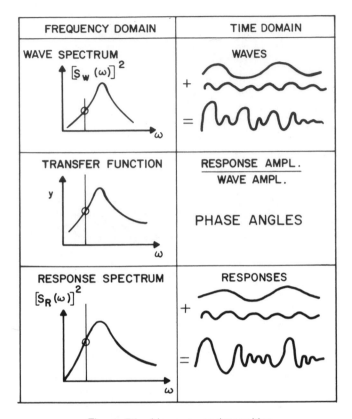

FREQUENCY DOMAIN	TIME DOMAIN
WAVE SPECTRUM $[S_w(\omega)]^2$	WAVES + =
TRANSFER FUNCTION y	RESPONSE AMPL. / WAVE AMPL. PHASE ANGLES
RESPONSE SPECTRUM $[S_R(\omega)]^2$	RESPONSES + =

Fig. 1–54 *Linear superimposition.*

shown in Figure 1-55 still yields. By applying H/2 instead of δ in Equation (20), we get:

$$P_S(H) = 1 - \exp.\left(-\frac{H^2}{4E}\right) \qquad (23)$$

The parameter, E, is related to the area beneath the wave spectrum curve by the following equation:

$$E = 2S_w^2 \qquad (24)$$

It may be easily shown that the following equation yields:

$$H_{1/3} \cong \sqrt{8E} = 4S_w \qquad (25)$$

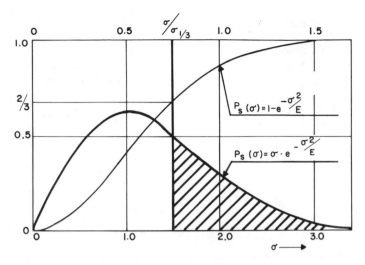

Fig. 1–55 *Rayleigh distribution function.*

The distribution function may then be written:

$$P_s(H) = 1 - \exp.\left[-2\left(\frac{H}{H_{1/3}}\right)^2\right] \qquad (26)$$

The most probable largest wave height during the time interval, t, may then be calculated as:

$$H_{max} = H_{1/3}\sqrt{0.5 \ln (^t/\overline{T})} \qquad (27)$$

The average apparent wave period, $\overline{T}$, is usually determined so as to give the most unfavorable response with due regard to obtaining reasonable wave steepness. This parameter is the most difficult to choose, and for the investigations performed so far it has been determined according to the above criteria.

According to additional investigations now being performed, it seems that the period chosen is not as critical for this procedure as it might be when using a design wave.

b) *Long Term Statistics*

It is evident that it is far from sufficient to know the statistical distribution of wave heights and response amplitudes at stationary conditions. How will the waves and the response amplitudes behave during long periods of time (months, years) when $H_{1/3}$ and $\overline{T}$ vary?

The long term, nonstationary response may be described as the sum of a large number of short term stationary processes. This technique is expressed by the following mathematical relationship:

$$P(\sigma) = \int_0^\infty P_s(\sigma) \cdot p(\sqrt{E}) \, d\sqrt{E} \qquad (28)$$

Based on a large amount of data, the proper distribution for E or $\sqrt{E}$ was found. This is the Weibull distribution function, and is defined as follows:

$$P(\sqrt{E}) = 1 - \exp.\left[-\left(\frac{\sqrt{E}}{A_1}\right)^M\right] \qquad (29)$$

Equation (28) was analyzed with long term distribution of $\sqrt{E}$ according to Equation (29). The result was found to be another Weibull distribution, defined as follows:

$$P(\sigma) = 1 - \exp.\left[-\left(\frac{B}{b}\right)^K\right] \qquad (30)$$

where:

$$B = \left(\frac{\sigma}{A_1}\right)^2 \qquad (31)$$

The parameters b and K are functions of the parameter M in Equation (29).

Thus a long term distribution of the response amplitude δ has been determined.

Some results concerning bending moment amidship for large tankers are shown in Figure 1-56.

It may also be shown that wave heights follow the long term Weibull distribution:

$$P(H_v) = 1 - \exp.\left[-\left(\frac{H_v - H_o}{H_c - H_o}\right)^G\right] \qquad (32)$$

Here $\dot{H}_v$ denotes visually observed wave height. Some

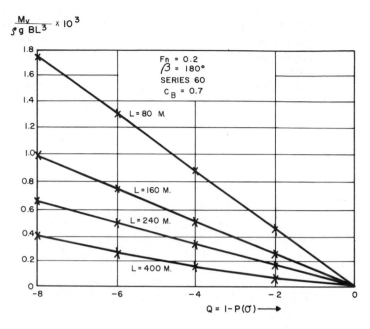

Fig. 1–56 *Calculated long term distributions of bending moment amidships.*

examples of this distribution are shown in Figure 1-57 ($H_o = O$).

By using an experimentally determined function to transfer visually observed wave heights to significant wave heights, the results shown in Figure 1-58 may be obtained. Further information of this principle is given by Nordenstrom.[11]

Design Wave Method

The design wave method is in principle applicable to both fixed and floating offshore structures. At present, however, it is mainly used for fixed offshore structures rather than floating drilling units. The height of the design wave (50 or 100 year wave) is determined by long term wave statistics as presented above, with results shown in Figures 1-57 and 1-58.

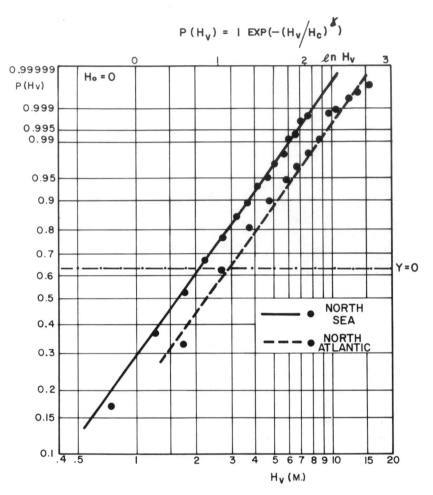

Fig. 1–57 *Weibull distributions of visual wave heights.*

In determining the total wave forces on fixed offshore structures, the force on long, slender, cylindrical elements of arbitrary cross-section form is of great interest to the designer.

1. *Long Slender Elements*
   The wave and current forces on these elements (for in-

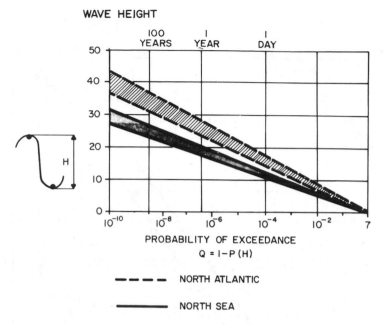

Fig. 1–58 *Predicted long term distributions of wave heights.*

stance, piles) may be reasonably well predicted by applying the Morison equation:

$$F_W = \frac{1}{2}\ \zeta \cdot g C_D \Delta L D V^2 \operatorname{Sin} a_1 + \frac{\pi}{4}\zeta \cdot C_M\ D^2 \cdot \Delta L \dot{V} \cdot \operatorname{Sin} \beta_1 \qquad (33)$$

The drag coefficient C_D ranges from 0.7 for circular cylinders to 1.5 for structural steel shapes. The added mass coefficient $C_m = C_M - 1$ ranges from 1.0 for circular cylinders to 2.0 for structural steel shapes.

The velocity term is seen to be identical in form to the wind force expression (see Equation 7) depending on the velocity squared. The current velocity is added vectorially to the water particle velocity in the wave in order to determine V.

The inertia force is proportional to the diameter squared and linear with respect to acceleration.

The range in the factors C_D and C_m indicates the importance of rounded shapes which produce less drag forces and of open construction with as little area exposed to the wave action as possible in order to reduce the forces due to wave and current on the structure.

The variation in C_D with Reynold's number R_E for circular cylinders is shown in Figure 1-59.[12] It should be emphasized that in the region for practical design, C_D should not be chosen less than 0.7.

The variation in C_m for cylinders with rectangular cross-sections is shown in Figure 1-60.

The force has been seen to be a function of velocity squared. Consequently, no linear relationship exists and no transfer function for the force or moment can be established as long as the drag forces are of significant influence (small diameter elements). The spectral analysis method in the form used for floating structures is not applicable to fixed structures when the drag forces are of significant influence to the overall loads. This is due to the nonlinearity of the drag forces. Yet no other statistical method for practical use involving nonlinear statistics exists. Thus, a design wave of specific height and period is used.

The water particle velocity and acceleration may be determined from wave theories such as Stoke's third or fifth

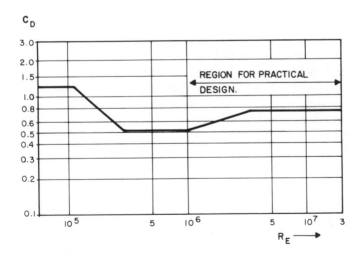

Fig. 1–59 *Drag coefficient versus R_E circular cylinder.*

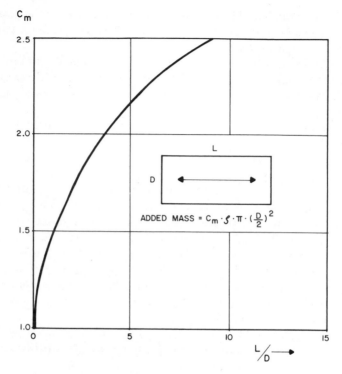

Fig. 1–60   *Added mass coefficient C_m vs L/D. (rectangular section).*

order, onoidal, linear theory, etc. These are all solutions to
the basic governing LaPlace equation. The differences in the
various theories are found in the number of terms in the
solution of the LaPlace equation and satisfaction of the
boundary conditions at the water surface and at the sea
bottom.

The degree of accuracy (and resulting complications) depends upon wave length, water depth, and ratio of wave
depth to wave length and wave height to water depth. This is
illustrated in Figure 1-61.

The velocities and accelerations of water particle velocities due to wave action vary with distance from the crest and
with depth, as shown in Figure 1-62. The particle velocities
are plotted for the mean water level. Note that the

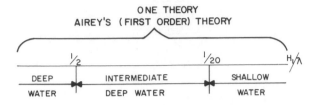

SMALL AMPLITUDE WAVES

ONE THEORY
AIREY'S (FIRST ORDER) THEORY

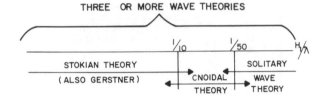

FINITE AMPLITUDE WAVES

THREE OR MORE WAVE THEORIES

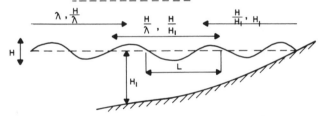

IMPORTANT PARAMETERS

Fig. 1–61 *Classification of amplitude waves.*

horizontal particle velocity is maximum at the crest of the wave while the horizontal acceleration is zero. The horizontal acceleration on the other hand is maximum at the wave node while the horizontal velocity is zero.

Figure 1-63 shows the path that the water particles travel during the passage of a wave.

The maximum force on any submerged member occurs when the resultant of the drag and inertia force is maximum. For large diamater elements when the inertia force is predominant, this will occur approximately at the point when the wave node passes the element.

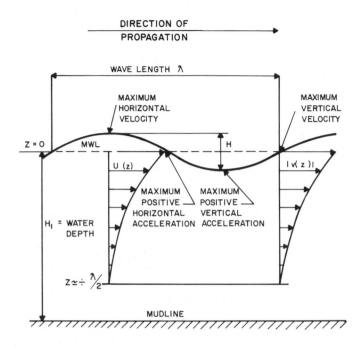

Fig. 1–62 *Variation of particle velocity with depth below sea surface.*

2. Large Volume Structures

Slender elements like those just considered do not influence the velocity and acceleration field around the structure. A large body (e.g., oil storage tank) will, however, reflect the waves producing disturbances to the velocity and acceleration field. An increase in velocities and accelerations around the body may result.

Consider for instance the structure shown in Figure 1-64. The forces on the towers may be calculated by applying the Morison's equation, but the water particle kinematics should be corrected for the presence of the caisson. This is illustrated in Figure 1-65. This effect may be approximately accounted for by applying the water depth above the caisson as total water depth in the calculation. Another more exact method is to calculate the water particle kinematics in the

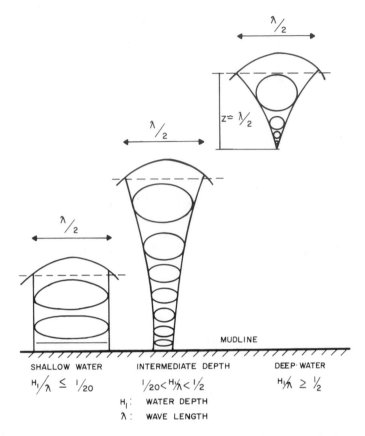

$z \cong \lambda/2$

$\lambda/2$

$\lambda/2$

$\lambda/2$

MUDLINE

SHALLOW WATER   INTERMEDIATE DEPTH   DEEP· WATER

$H_1/\lambda \leq 1/20$ $1/20 < H_1/\lambda < 1/2$ $H_1/\lambda \geq 1/2$

H_1 :   WATER DEPTH

λ :   WAVE LENGTH

Fig. 1–63 *Water particle trajectories.*

fluid by means of diffraction theory, which takes into account the reflection of waves from the caisson.

The forces and moments on the caisson may be calculated in one of two ways:

1. By integrating the pressure resulting from the incoming wave and applying experimentally determined amplification factors to forces and moment.
2. By diffraction theory which takes into account the reflection of waves. This technique uses a large number of sinks

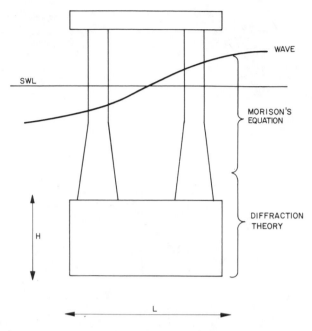

Fig. 1–64   *Force calculation.*

and sources distributed over the average wetted surface of the body.

The last method is preferred as the form of the structure could be more exactly taken into account than by the first method. It should, however, be mentioned that when using the diffraction theory, analytical solutions are not available for caisson forms other than cylindrical. Numerical solutions by means of computer are possible, and DnV has developed such a computer program for fixed and floating large bodies of any form 13. Some results are shown in Figure 1-66.

The maximum forces and moments on an offshore structure such as that shown in Figure 1-64 will occur when the wave position is as indicated in Figure 1-67. When drag forces are significant, the maximum horizontal force and

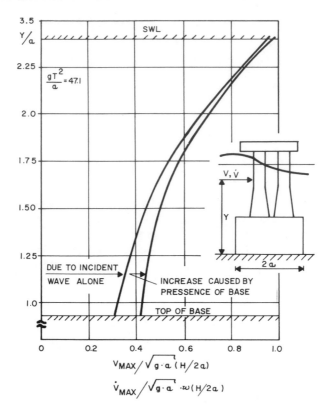

Fig. 1–65 *Velocity and acceleration distribution above base.*

overturning moment on the whole structure will occur at a wave position between the two shown. The exact wave position depends on the ratio of drag force to inertia force and the magnitude of the horizontal force on the caisson alone.

In order to minimize the total overturning moment and the horizontal force, the ratio of caisson height to caisson length should be chosen as small as possible. When this ratio is small enough, the caisson will have a stabilizing effect due to the moment contribution from the vertical wave force. This effect is illustrated in Figure 1-68.

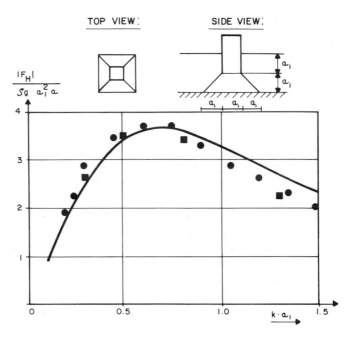

Fig. 1–66   *Oscillating wave force.*

Wave Spectra

The wave spectrum most generally used by engineers in the last 10 years is the Pierson-Moskowitz spectrum proposed in 1964.[14] A new spectrum which has not yet been extensively applied by engineers is the Jonswap spectrum. This spectrum is a result of the Joint North Sea Wave Observation Project—a comprehensive international experiment undertaken in the North Sea off the Island of Sylt.[15] The shape of the Jonswap spectrum is very different from that of the Pierson-Moskowitz

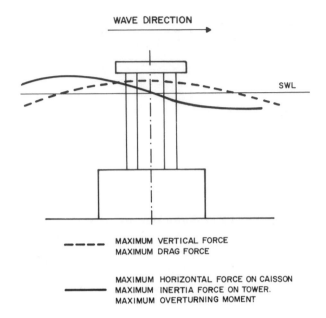

WAVE DIRECTION

SWL

- - - - - MAXIMUM VERTICAL FORCE
MAXIMUM DRAG FORCE

MAXIMUM HORIZONTAL FORCE ON CAISSON
MAXIMUM INERTIA FORCE ON TOWER.
MAXIMUM OVERTURNING MOMENT

Fig. 1–67   *Wave profiles at maximum loads.*

spectrum. The major difference between these two spectra is that the Jonswap spectrum is more peaked. When the original Pierson-Moskowitz spectrum is used for comparison, the Jonswap spectrum also contains more energy (Figure 1-69).

These two spectra cannot, however, represent the same sea state as the significant wave height for each is different due to the difference in the area beneath the spectra curves. Generally, it may be stated that when representing the same sea state, the Pierson-Moskowitz spectrum distributes the energy over a wider range of frequencies than does the Jonswap spectrum, provided that the areas beneath the spectra curves are equal (Figure 1-70).

The Jonswap spectrum and the original Pierson-Moskowitz wave spectrum with frequency $f(h_z)$ as a parameter may be expressed as:

$$E(f) = a_2 \cdot g^2 \cdot (2\pi)^{-4} f^{-5} \exp.\left[-\frac{5}{4}\left(\frac{f}{f_m}\right)^{-4}\right] \cdot \gamma^{\exp.\left[\frac{-(f-f_m)^2}{2\sigma_1^2 \cdot f_m^2}\right]}$$

$$a_2 = 0.008 \tag{34}$$

$$\sigma_1 = \begin{array}{l} 0.07 \text{ for } f \leq f_m \\ 0.09 \text{ for } f > f_m \end{array}$$

f_m = Peak frequency

γ = peakedness parameter

γ = 1 (Original Pierson-Moskowitz) ⎫
                                                    ⎬ See Figure 1-69
γ = 3.3 (Average Jonswap)   ⎭

The modified Pierson-Moskowitz wave spectrum may be expressed as:

$$E(f) = \beta_2 \frac{H_{1/3}^2}{\pi \cdot f_m}\left(\beta_2 \frac{f}{f_m}\right)^{-5} \exp.\left[-\frac{1}{\pi}\left(\beta_3 \frac{f}{f_m}\right)^{-4}\right] \tag{35}$$

$$\beta_2 = 0.178 \;⎫$$
$$\qquad\qquad\quad ⎬ \text{ See Figure 1-70}$$
$$\beta_3 = 0.71 \;\;⎭$$

Equation (35) is obtained from Equation (15) applying the following relationships:

$$\omega = 2\pi f \tag{36}$$

$$T_m = 1.408\ \overline{T}\ \text{(Valid only for Pierson-Moskowitz Spectrum)} \tag{37}$$

$$f = \frac{1}{T} \tag{38}$$

Which of these two spectra gives the best representation of reality depends obviously on the actual location. Analysis of

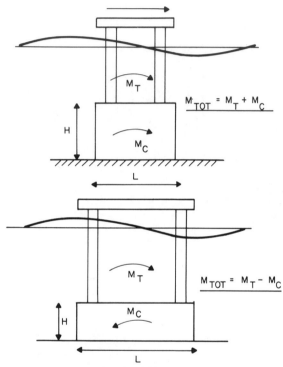

M_T : MOMENT ON TOWER

M_C : MOMENT ON CAISSON

Fig. 1–68   *Effect of L/H ratio.*

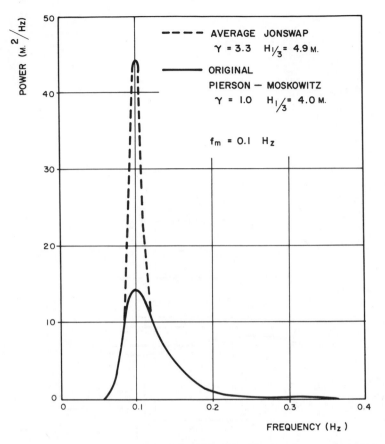

Fig. 1–69 *Jonswap and original Pierson-Moskowitz wave spectra.*

data recorded off the coast of Northern Norway do, however, strongly indicate that the Jonswap spectrum is the most representative for the North Sea and adjacent waters.[15]

APPLICATION TO FIXED OFFSHORE STRUCTURES

The wave loads on fixed offshore structures are normally evaluated by means of one of two basic methods:
a) Design wave method

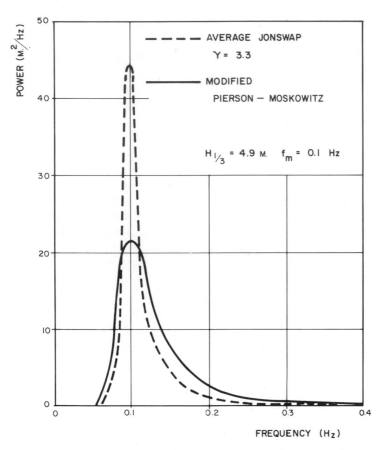

Fig. 1–70 *Jonswap and modified Pierson-Moskowitz wave spectra.*

b) Spectral analysis method

The method to be used should be considered in each specific case with due regard to the design and purpose of calculation.

Design Wave Method

Until now, most analysis regarding wave loads on fixed offshore structures has been based on a regular, sinusoidal design wave of given height and period. The wave height to be applied is evaluated from wave statistics for the area of interest, taken as the height of the 50 or 100 year wave, which may be determined with a reasonable degree of accuracy. The wave period is, however, more complicated to determine.

Normally, the principle that the period should correspond to the worst case loading is adopted as a guideline. This implies that the loads have to be calculated for several wave periods in order to establish the most severe case. A problem may appear, however, when considering total horizontal force on a large volume structure such as the one shown in Figure 1-71. The force is often found to increase continuously with increasing wave period towards a peak value far above the T = 20 seconds generally adopted as the upper limit for the period. The transfer function for horizontal force on the structure in Figure 1-71 is shown in Figure 1-72. The force is seen to attain its maximum magnitude at a wave period of about 30 seconds.

The difficulties of establishing a design criterion for the wave period are in this case obvious. To apply the maximum force with no regard for the period would lead to unrealistic results in many cases. Further investigation is necessary to clarify this problem, but until better knowledge of the subject is obtained, a 20 second limit derived from energy considerations seems appropriate.

Another effect which is important to consider is that maximum force and overturning moment may be obtained for different wave periods. This implies that the design wave period giving the most unfavorable conditions ought to be chosen separately for each response. Generally, it may be stated that the overturning moment will achieve its maximum for lower periods than the horizontal force. The transfer function for overturning moment on the structure in Figure 1-71 is shown in Figure 1-73.

Spectral Analysis Method

Spectral analysis of a fixed offshore structure can be justified only if the transfer functions for the responses can be established with a reasonable degree of accuracy. This implies that the nonlinear loads (for instance drag forces) have to be small com-

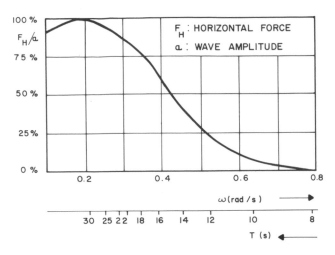

Fig. 1–72   *Transfer function-horizontal force.*

pared to the linear loads (for instance inertia forces). If not, no well defined transfer function exists, and any result whatsoever could be obtained depending on the transfer function chosen. Linearization of nonlinear loads should be done with care and with due regard to their importance to the overall loads.

It should be emphasized that, even if a transfer function is established, the conditions may be very different from those applying to ships and other floating structures. The main differences are related to the shape of the transfer functions. Ship responses can normally be described by transfer functions having a well defined peak value at relatively low wave periods while this is not necessarily true for fixed offshore structures. The horizontal force on the structure in Figure 1-71 may be described by the transfer function in Figure 1-72, which is seen to increase continuously towards a peak value at about T = 30 seconds.

Combining this transfer function with the wave spectrum will give increasing values with increasing peak period of the spectrum far above T = 20 seconds. Consequently, careful consideration must be given to the location of the wave spectrum along the frequency axis when performing short term analysis.

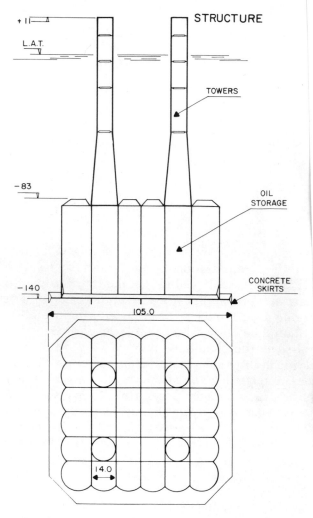

Fig. 1–71 *Total horizontal force on large volume structure.*

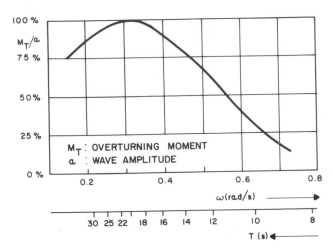

Fig. 1–73 *Transfer function-overturning moment.*

The choice of the period is to a great extent avoided, however, when carrying out a long term prediction according to the method normally used in ship design.[11] The long term prediction is obtained by summing up a great number of short term statistically stationary conditions, each combined with a specified probability of occurrence. Even if the short term conditions obtained by locating the peak period of the spectrum above 20 seconds give unrealistic results, they will be associated with low probabilities of occurrence and consequently have little influence on the long term predictions.

The problems involved in spectral analysis are seen to be many, especially when only short term predictions are desired. Further investigatios are definitely needed to establish reasonable design criteria. Meanwhile, it seems appropriate to apply the 20 second limit to the peak period of the spectrum even in this case.

Conclusions—Fixed Structures

The analysis of wave forces may be based on the design wave method or on the spectral analysis method. Spectral analysis is in principle considered to give more reliable results than the design wave method, provided that the transfer functions for the responses in question can be established with a reasonable degree of accuracy.

Long term analysis is generally perferred. If, however, only short term analysis is carried out, the location of the spectrum along the frequency axis should be properly justified.

In the cases where the above assumptions are not fulfilled, the design wave method should be applied.

FUTURE ASPECTS

It has been attempted here to give a view of the methods that are commonly used today in calculating wind, wave, and current forces on offshore structures. As the reader has most certainly noticed, there are certain problems involved in applying the spectral analysis method to fixed offshore structures due to nonlinearities in the loads, which may be of great influence. Methods involving nonlinear statistics for practical application are not yet available.

As the forces and moments are statistical in nature, however, the need for statistical methods which take the nonlinearities into account is evident. It is hoped that strong efforts in clarifying this field will be made in the near future, and Det norske Veritas hopes to participate actively in this research in the near future.

Nomenclature

A   Projected area of a member taken as the projection on a plane normal to the direction of the considered force.

A_1 Parameter of $P(\sqrt{E})$.

a   Elementary wave amplitude.

B $\left(\dfrac{\sigma}{A_1}\right)^2$

b   Parameter of $P(\sigma)$.

C	Shape coefficient (3-dimensional flow).
C_∞	Shape coefficient (2-dimensional flow).
C_D	Drag coefficient.
C_H	Height coefficient.
C_M	$= 1 + C_m$ (inertia coefficient).
C_m	Added mass coefficient.
D	Diameter.
d	Cross sectional dimension normal to the direction of the considered force.
E	Parameter of the short term Rayleigh distribution.
E(f)	Wave spectrum as a function of frequency f (h_z).
F_H	Total horizontal force on a large volume structure.
F_W	Wind force acting perpendicular to an area, A.
F_Y	Wind force acting in direction of the wind.
f(a)	Directionality function.
f	Frequency (h_z).
f_m	Peak frequency of the wave spectrum (h_z).
g	Acceleration due to gravity.
H	Individual crest-to-trough wave height.
H_1	Still water depth.
H_c, H_0, G	Parameter of $P(H_v)$.
H_v	Visually observed wave height.
$H_{1/3}$	Significant wave height.
$H_{max.}$	Most probable largest wave height.
i	Wave frequency number.
K	Parameter of $P(\sigma)$.
k	Wave number ($2\pi/r$).
L	Length of member.
ΔL	Length of element or portion of structure considered.
l	Wave train number.
M	Parameter of $P(\sqrt{E})$.
M_T	Total overturning moment on a large volume structure.
$P(\sigma)$	The long term Weibull distribution of the response amplitude σ
$Ps(\sigma)$	The short term Rayleigh distribution of the response amplitude σ
$P(\sqrt{E})$	The long term Weibull distribution of $\sqrt{E}$
$p(\sqrt{E})$	$\dfrac{d\,P\,(\sqrt{E}\,)}{d\,P\,\sqrt{E}}$
P(H)	Long term Weibull distribution of individual wave heights.
$P(H_v)$	Long term Weibull distribution of visually observed wave heights.

R_E	$\dfrac{V \cdot D}{\gamma}$	Reynold's number.
$S_R{}^2$		Area beneath the response spectrum curve.
$S_W{}^2$		Area beneath the wave spectrum curve.
$S_R(\omega)^2$		Response spectrum.
$S_W(\omega)^2$		Two-dimensional wave spectrum.
$S_W(\omega, a)^2$		Three-dimensional wave spectrum.
T		Zero-uncrossing period.
$\overline{T}$		Average apparent zero-upcrossing period.
T_m		Peak period of the wave spectrum.
T_R		Average apparent zero-upcrossing period of the response amplitude σ.
t		Time interval.
V		Water particle velocity due to wave and current.
$\dot{V}$		Water particle acceleration due to wave.
V_{cy}		Total current velocity y meters above sea bottom.
V_T		Tide-induced current velocity at sea surface.
V_W		Wind-induced current velocity at sea surface.
V_z		Sustained wind velocity z meters above sea surface.
$(V_z)_G$		Gust wind velocity z meters above sea surface.
V_{10}		Sustained wind velocity at 10 meters above sea surface.
V_{10}		Sustained wind velocity at 10 meters above sea surface acting normal to an area, A.
Y		Transfer function.
y		Distance above sea bottom.
z		Distance above sea surface.
a		Angle between main direction of propagation of short crested wave system and the elementary long crested wave systems.
a_1		Angle between velocity vector and axis (or surface) of member.
β		Heading angle.
β_1		Angle between acceleration vector and axis (or surface) of member.
λ		Wave length.
ω		Circular wave frequency.
σ		Elementary response amplitude.
ζ		Phase angle.
ξ		Density of fluid.
γ_1		Kinematic viscosity.
γ		Peakedness parameter.

REFERENCES

1. Salvesen, N., Tuck, E. O., and Faltinsen, O.: "Ship Motions and Sea Loads," Det norske Veritas Publication No. 75, March 1971.
2. Nordenstrom, N., Faltinsen, O., and Pedersen, B.: "Prediction of Wave-Induced Motions and Loads for Catamarans," Det norske Veritas Publication No. 77, September 1971.
3. Pedersen, B., Egeland, O., and Langfeldt, J. N.: "Calculation of Long Term Values for Motions and Structural Response of Mobile Drilling Rigs," Offshore Technology Conference, Reprints Paper No. 1881, Houston, Texas, April 29-May 2, 1973.
4. Olsen, O. A.: "Wind Statistics for the North Sea and the Norwegian Ocean," Det norske Veritas Report No. 74-53-D.
5. Davenport, A. G.: "Wind Loads on Structures," National Research Council, Division of Building Research, Technical Paper No. 88, Ottawa, March 1960.
6. "Rules for the Design, Construction and Inspection of Fixed Offshore Structures," Det norske Veritas, Oslo, 1974.
7. "Rules for the Construction and Classification of Mobile Offshore Units," Det norske Veritas, Oslo, 1973.
8. Tung, Chi Chao, and Huang, Norden: "Combined Effects of Current and Waves on Fluid Force," *Ocean Engineering,* Volume 2, 1973.
9. Langfeldt, J. N., Egeland, O. and Gran, S.: "NV407. Motions and Loads for Drilling Platforms with Pontoons," User's Manual, Det norske Veritas Report No. 72-13-S.
10. Faltinsen, O.: "Computer Program Specification NV417. Second Edition. Wave-Induced Ship Motions and Loads. Six Degrees of Freedom," Det norske Veritas Report No. 74-19-S.
11. Nordenstrom, N.: "A Method to Predict Long-Term Distributions of Waves and Wave-Induced Motions on Loads on Ships and other Floating Structures," Det norske Veritas Publication No. 81, April 1973.
12. Olsen, O. A.: "Investigation of Drag-Coefficient Based on Published Literature," Det norske Veritas Report No. 74-2-S.
13. Faltinsen, O. M., and Loken, A. E.: "NV459. Wave Forces on Large Objects of Arbitrary Form," User's Manual, Det norske Veritas Report No. 74-13-S.
14. Pierson, W. J., and Moskowitz, L.: "A Proposed Spectra Form for Fully Developed Wind Seas Based on the Similarity Theory of S. A. Kitaigorodskii," *Journal of Geophys. Res.,* V. 69 (24), 1964.
15. Hasselmann, K., et. al.: "Measurements of Wind-Wave Growth and Swell during the Joint North Sea Wave Project," *Deutsches Hydrographisches Zeitschrift,* Nr. 12, 1973.

4

Spread Mooring Systems

Mark A. Childers
ODECO, Inc.

Purpose and Basic Design Criteria

In floating drilling, the purpose of the mooring system is to maintain the drilling vessel within certain horizontal limits of the centerline of the well so that drilling operations can be successfully carried out. This horizontal excursion during actual drilling operations is usually held to a maximum of 5 to 6% of water depth; however, most drilling operations are carried out within 2 to 3%. These limits are controlled by the subsea drilling equipment such as stresses in the marine riser, angle of the lower ball joint, and the nature of the drilling operation.

During nonoperating times when the marine riser is still connected to the blowout preventer (BOP) stack, the mooring system is usually designed to maintain the drilling vessel within approximately eight to ten percent of water depth. Once again, these limits are controlled by the subsea drilling equipment. During maximum or survival conditions when the marine riser is disconnected from the BOP stack, the amount of excursion off the hole is secondary to relieving high mooring line tensions.

Except during maximum conditions, it is necessary to think of the mooring and subsea system as working together. Based primarily on field experience and supported by analytical analysis, Table 1-3 gives a summary of the generally accepted design and operating criteria for mooring operations as generally discussed above.

By far, the most common type of mooring system is the spread mooring system which historically has consisted of from one to twelve mooring lines. Dynamic positioning, which has no physical connection to the ocean bottom other than the subsea drilling equipment, is also used on a limited number of vessels. At present, the station-keeping ability of the dynamically positioned ships is considerably less than a properly designed spread mooring system. The primary fortes of dynamic position mooring are ultra-deep water, high mobility and rapid well abandonment.

Spread mooring systems have been successfully used in 1,7500 foot water depths and extensive analytical analysis indicates that they can be very successful in over 3,000 feet. However, there are some water depths, probably over 4,000 feet, where the spread mooring concept becomes impractical from an economic as well as a station-keeping standpoint.

Spread Mooring Patterns

There are many possible arrangements for spacing the mooring lines around the drilling vessel in a spread mooring pattern. The reason for the large number of different patterns stems from an effort to gain the maximum restoring force from the mooring system in combination with the environmental resistance or loading characteristics of the vessel. As may be expected, the environmental loading characteristics of ships and semisubmersibles vary from the bow to a beam environmental attack angle. Most ships exhibit a maximum wind, wave, and current load on the beam, whereas most semisubmersibles exhibit a maximum loading condition in the vicinity of 45° off the bow or stern.

In general, there are two types of mooring patterns which can be used with any one particular type of vessel. The first is the "Omnidirectional Attack" pattern, which is arranged to take

TABLE 1-3

Design Parameters And Conditions Used For Spread Mooring Analysis

Nominal Operation Designation	Mooring Line		Maximum Vessel Offset (%)	Marine Riser Situation			Operation
	Maximum Tension	Leeward Line Slacking Policy		Condition	Max. Lower Ball Angle (Degrees)	Mud	
Normal Drilling	⅓ Break	Nominal	Approx. 3	Connected	4	Drilling Mud	Drilling ahead; running casing, BOP, test tools & performing all normal operations.
Drilling	⅓ Break	Nominal	6	Connected	Under 10	Drilling Mud	Preparing to wait on weather, hoisting, pulling riser, setting cement or barite plugs & critical drilling operations.
Standby	⅓-½ Break	Equivalent to at least 2 leeward lines completely slackened	10	Connected	10	Displaced w/sea water as needed	No drilling operations with riser ready to be disconnected at moments notice. Waiting on weather.
Survival or Maximum	½ Break or anchor slippage	Equivalent to at least 2 leeward line completely slackened	Unrestricted	Disconnected	—	—	No operations of any kind except possible mooring line manipulation. Rig may not be manned.

environmental loads from any attack angle (0 to 360°). In areas such as the North Sea and the Gulf of Alaska where maximum winds and waves can arrive from any direction, the omnidirectional attack pattern is required.

The second type of pattern is the "Unidirectional Attack" pattern in which there is a strong prevailing environmental direction. The unidirectional attack pattern is stronger in one direction, usually the bow, than the omnidirectional pattern, but it is weaker in a direction approximately 90° from the strongest direction. The unidirectional attack pattern is advisable in such areas as the mouth of the Amazon River off Brazil, Cook Inlet in Alaska, and certain areas off South Africa where a strong prevailing current always exists in one direction or 180° from that direction.

Deployment of the proper mooring pattern is a very important factor in reducing mooring line loads and staying within desired horizontal displacement tolerances. A proper spread mooring pattern should coincide with the environmental wind, wave, and current loading characteristics of the drilling vessel.

For an "Omnidirectional Attack" pattern, for example, ships will "stack" or align the pattern to be strongest in the beam direction. Figure 1-74 shows six of the most commonly used types of "Omnidirectional Attack" patterns. Figure 1-74(a) shows a symmetrical nine line pattern which is used by the SEDCO 135 and ODECO "Ocean Driller" class. The symmetrical eight line system shown in Figure 1-74(b) is used on such vessels as The Offshore Company's Discoverer Class and has been used on the Santa Fe Bluewater 2.

Figure 1-74(c) shows the pattern used by the Pentagone 81 semisubmersible and Figure 1-74(d) shows the pattern used by a number of Global Marine vessels, such as the Glomar Grand Isle. Figure 1-74(e) shows the pattern being used on the Briet Engineering design semisubmersible and some shipshape vessels.

Perhaps the most common mooring pattern is that shown in Figure 1-74(f) which was initially used on the LST Drilling Tenders in the Gulf of Mexico but is now used on vessels such as the ODECO "Ocean Victory" and "Ocean Queen" class.

The spread mooring system is very inefficient in that less than

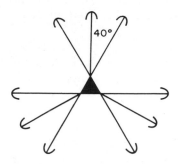

a) SYMMETRIC NINE LINE

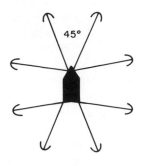

b) SYMMETRIC EIGHT LINE

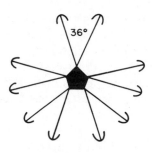

c) SYMMETRIC TEN LINE

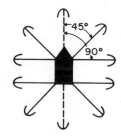

d) 45° – 90° EIGHT LINE &
   45° – 90° TEN LINE

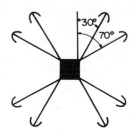

e) 30° – 70° EIGHT LINE

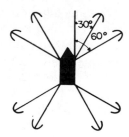

f) 30° – 60° EIGHT LINE

Fig. 1–74 *Typical spread mooring patterns.*

half of the mooring lines contribute to holding the vessel on location at any one time. Of this half, usually only one or two lines supply most of the restoring force. In fact, if the leeward lines are not manually slackened during severe conditions, they actually draw the vessel off location and contribute to higher mooring windward line tensions.

Anchors

Before the station-keeping ability of any spread mooring system can be understood or analyzed, the characteristics of the anchors, which are the main holding force of the mooring system, must be discussed. In floating drilling, the anchors most commonly used are designated "dynamic anchors" because they increase their holding power with increased horizontal pull providing there is no vertical uplifting force. To prevent an uplifting force on the anchor, sufficient mooring line must be deployed such that the catenary is tangent to the ocean floor at or before the anchor.

Tests conducted by the U.S. Navy indicate that only a nominal decrease in holding power will result with a 6° angle with horizontal; however, a significant loss in holding power will result if the angle increases to 12°. For design purposes, enough mooring line should be outboard of the lower fairleader *after pretensioning* such that the mooring line is tangent to the ocean bottom before the anchor.

Basically, there are three types of anchors used, the most common being the Light Weight Type (LWT), with the other two being the Stato type and Danforth. Figure 1-75 shows the nomenclature used with these types of anchors. The Stato type comes in a fabricated and cast version. Shop tests as well as field

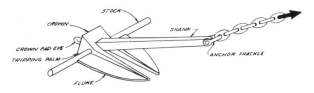

Fig. 1–75 *Dynamic anchor nomenclature.*

experience indicate that the cast version with a trade name of Moorfast or Offdrill is structurally stronger and more durable.

All three types of anchors perform approximately the same in sand and clay. Maximum holding power is obtained when the flute angle is set at approximately 30° in sand and/or clay and 50° in soft mud. The Stato type anchor has shown superior tripping and "digging in" characteristics in soft bottoms such as the Gulf of Mexico. Tests conducted by the U.S. Navy have shown that along with soil conditions and other factors, maximum holding power varies with anchor weight.

For anchors in the 10,000 to 15,000 lb. range, a holding power of approximately 12 to 17 times their dry weight can be expected. In the range of 30,000 lb. anchors, this ratio decreases to approximately 8 to 11 times the anchor's dry weight. For anchors in the 45,000 lb. range, this factor decreases below 10. Figure 1-76 shows a log-log plot of anchor holding power vs. anchor dry weight for different soil conditions. Though the plot results in a straight line, as determined by small anchors and extrapolated to bigger anchors, some tests indicate that holding power decreases faster than indicated with increasing anchor size.

If poor holding power is experienced or expected, "piggybacking" the primary anchor in series with an additional anchor may be necessary. Before this is done in actual field practice, the anchor should be allowed to "soak" or remain relatively unloaded for a number of hours to allow the soil to consolidate and regain some of its sheer strength. Often this will result in the anchor producing the desired holding power. In any event, if "piggybacking" is necessary, calculations should be done to see that additional tension does not cause a vertical lifting force on any of the anchors.

Anchors should be properly "preloaded" in correspondence to the highest mooring line load expected, which may be as much as one-half the mooring line's rated break strength. At a minimum, the anchor should be preloaded to one-third the rated break strength, especially if design criteria such as those outlined in Table 1-3 are used.

In areas where anchor holding power is unknown, tests have shown that small anchors (approximately 3,000 lbs.) can give a

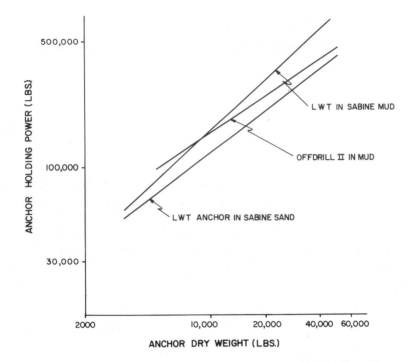

Fig. 1–76   *Anchor holding power vs anchor dry weight for LWT and Stato type (Offdrill II) anchors.*

reasonably good indication of large anchor holding power. Where dynamic anchors do not perform acceptably, some contractors and operators have used piled type anchors with limited success. However, they are extremely expensive and time consuming to install.

Explosive imbedding type anchors have never been used in floating drilling operations due to their limited holding power. However, in areas where dynamic type anchors do not perform properly, they show promise of satisfying a need when fully developed.

Mooring Lines

Mooring lines consisting of wire rope. chain, or a combination of the two have been successfully used in floating drilling.

Recently, synthetic fiber rope has been considered for mooring lines in combination with chain at the anchor end; however, no rigs have used or are currently using this type of mooring line. The selection of wire rope, chain, or a combination thereof for mooring lines rests on several factors which include expected mooring line loads, water depth, handling equipment, economics, and space for deck facilities onboard the rig and anchor handling boat.

The size, strength, and length of line depend upon the size and shape of the vessel, the working water depth, the expected environmental loading conditions, and the allowable horizontal vessel displacement as controlled by the subsea drilling equipment.

In general, for a given breaking strength, wire rope provides more restoring force than chain, particularly in deep water (1,000 feet or more) and large sizes (3 inches and more). Figure 1-77 shows single line tension of 3½ inch wire rope or 3¼ inch chain (both have approximately the same break strength) versus percentage of increased restoring force (horizontal force component) of 3½ inch wire rope over 3¼ inch chain as a function of

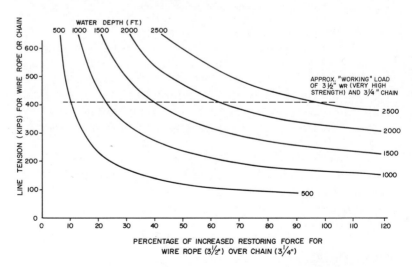

Fig. 1–77 *Percentage of increased restoring force for wire rope (3½ in.) over chain (3¼ in.).*

water depth. As seen, the advantage of wire rope over chain increases with increasing water depth and decreasing line tension. Solely from a restoring force standpoint, wire rope has a clear advantage over chain in deep water. However, restoring force, though probably one of the most important factors, is not the only factor that must be taken into account.

Figure 1-78 shows single line tension versus vessel displacement in the same plane for 3¼ inch chain and 3½ inch wire rope in 250 foot water depths. The difference in initial tension (pretension) at zero offset is controlled by the one-third rated break strength criteria at 6% displacement (15 feet).

In general, model tests and vessel motion simulators indicate that the mooring system has an almost negligible effect on vessel motions such as surge, sway, heave, etc. Therefore, for an as-

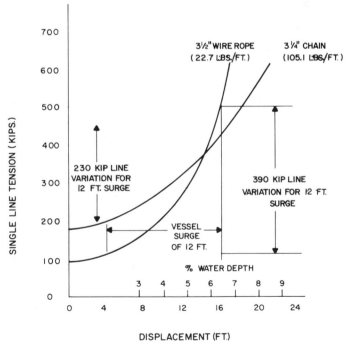

Fig. 1–78 *Line tension vs displacement for chain and wire rope in 250 ft. of water depth.*

sumed 12 feet of total vessel horizontal oscillation, Figure 1-78 shows that wire rope will have a peak to minimum line load spread of 390 kips. This large fluctuation in line tension will lead to accelerated fatigue in the mooring line. In addition, for this case increased vessel motion could easily escalate line tension to and over the break strength of the wire rope. In comparison, the fluctuation of the 3¼ inch chain is an undesirable 230 kips; however, additional surge will not readily break the chain.

Figure 1-79 shows 3½ inch wire rope and 3¼ inch chain single line tensions versus displacement in 1,000 foot water depths. For the same 12 feet of surge, which is now 1.2% of water depth instead of 4.8%, considerably less total line tension fluctuation occurs. Therefore, based on the above discussion, extensive field experience, and other factors to be discussed later, chain systems are in general terms better for relatively shallow

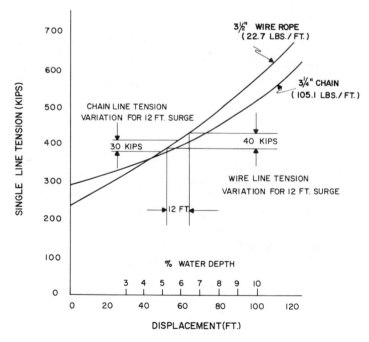

Fig. 1–79   *Line tension vs displacement for chain and wire rope in 1,000 ft. of water depth.*

water (approximately 600 feet). In fact, chain has shown its
durability and versatility in such operations as the Gulf of Mexico, where chain life may exceed ten years. Unfortunately, in
rough weather environments such as the North Sea, chain life
can be as little as four years.

In recent years, there has been a need to extend the water
depth capability of spread mooring systems. Presently, there are
a number of rigs which have spread mooring capability up to
3,000 to 3,500 ft. water depths. This is accomplished by using a
combination of wire rope (on the inboard side) and chain (on the
outboard side) such that significant restoring forces with
minimum total line length can be achieved. The combination
wire rope chain system (CWMS) can generate large restoring
forces in very deep water, at relatively low pretensions and with
very little mooring line manipulation to stay within desirable
offsets from the well, with low anchor handling boat shaft horse
power requirement to deploy the system, and other attractive
features. ODECO's "Ocean Ranger" class can presently be set up
to operate in 3,500 ft. of water via 5,600 ft. of 3½ in. extra high
strength wire rope and 3,000 ft. of Oil Rig Quality 3¼ in. chain.
Five other semisubmersibles are in operation and/or construction which will have the capability to operate in approximately
2,000 ft. of water. The limiting factor on water depth for the
CWMS system is more economics than station-keeping ability.
Operational considerations such as deployment and retrieving
are being developed which will significantly reduce the amount
of or elimination of pendant wire and surface buoys. Perhaps the
most significant advantage of the CWMS system is its use on
production floating platforms.

Salt water fatigue adversely affects the life of the expensive
mooring line. Field experience has shown that the maximum
"working" load of mooring line should be one-third the rated
break strength or less. In extremely rare situations, this load may
be allowed to reach one-half break strength; however, this situation is based more on total economics than on proper treatment
of the mooring line. The higher the load (especially over one-
third the break strength) and the larger the spread between the
low and high tensions during cycling, the shorter the life.

Vessel motion, excluding line harmonics, dictates the spread between the high and low load during cycling; therefore, semisubmersibles will in general have longer mooring line life since they have less motion than ships. In any event, from a fatigue (and thus life) standpoint, mooring line tensions between one-third and one-half break strength or more will, based solely on field experience, substantially reduce the normal operating life of the mooring lines.

The amount of mooring line which must be outboard of the lower fairlead after pretension is controlled by water depth and the maximum anticipated line tension. Since it is most desirable to have the mooring line catenary tangent to the ocean floor before or at the anchor, a single line catenary analysis is necessary. Figure 1-80 shows line tension versus suspended or catenary length from the lower fairlead to the ocean bottom tangent point for 3 inch Oil Rig Quality chain (89.3 lbs. per ft. dry weight) as a function of water depth.

By selecting the maximum anticipated tension, a cross plot of Figure 1-80 can be made. Figure 1-81 shows the required amount of 3 inch chain that must be outboard after pretension-

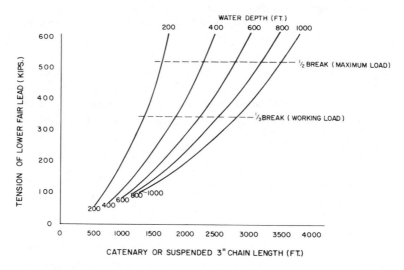

Fig. 1–80   *Tension vs catenary length as a function of water depth for 3 in. Oil Rig Quality chain (89.3 lb./ft.).*

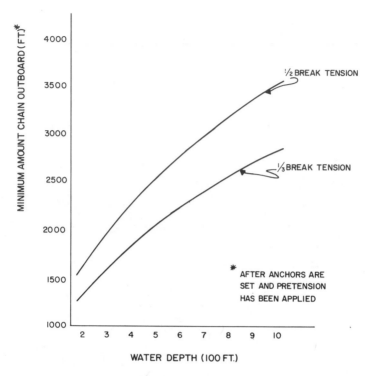

Fig. 1–81   *Amount of chain necessary to be outboard of lower fairlead after pretensioning*. (3 in. chain-89.3 lb./ft.).*

ing versus water depth for one-third and one-half break tension. Of interest is the fact that as water depth increases, the scope (mooring line length divided by water depth) decreases as shown in Figure 1-82. As will be discussed later, the rig must carry and thus deploy more chain than determined by analysis as in Figures 1-80 and 1-81, since additional line length is needed to set the anchor (50 to 150 feet), manipulate the mooring line during severe weather, and pretension the mooring lines.

Chain and Wire Rope Construction

Wire rope for mooring lines must be carefully selected as to strength (type of steel), size, lubricant, construction, internal

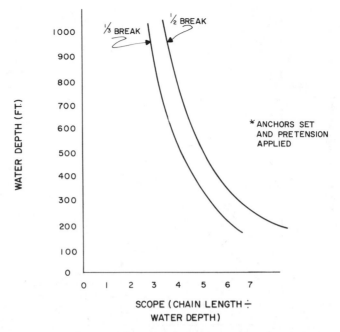

Fig. 1–82 *Required scope outboard of lower fairlead after pretensioning* (3 in. chain-89.3 lb./ft.) vs water depth as a function of maximum anticipated line tention.*

core, etc. Wire rope diameters up to 3½ inch have been used for mooring with the more common sizes being 2¼ to 3 inches. The wire rope should have an independent wire rope core (IWRC) center for crushing resistance and a zinc coating for corrosion protection since some of the wire rope will always be in the splash zone. Regular lay, usual right regular lay (RRL), is recommended since it has lower torque characteristics than lang lay. Six strand wire rope is recommended with each strand having 36, 41, or 49 wires depending on the size, strength, and type of steel.

Wire rope weighs approximately one-fifth as much as chain in water and currently costs approximately half as much as chain on a per foot basis. Due to wire rope's lighter weight, more line deployment is required to develop a tangent catenary at or

Fig. 1–83   *Chain nomenclature (flash welded).*

before the anchor. One of the big disadvantages of wire rope is that if severely damaged other than at or near the ends, the entire length must be discarded. However, wire rope performance on some 50 rigs indicates that this type of damage does not occur frequently. The in-water buoyancy factor for wire rope is 0.83 0.85 compared to 0.87 for chain.

There are two types of stud link chain which are currently being used in floating drilling—namely, flash welded chain and dialock chain. Figure 1-83 shows chain nomenclature and an example of flash welded chain. Flash welded chain is constructed of rolled bars which are then cut into link blanks, resistance heated, partially bent and inserted into the previous length, then completely enclosed into the overall configuration and flash welded. The studs are held into place by friction, and are not welded unless specified. It is highly recommended that at least one side of the stud be welded (side opposite the flash weld). Dialock and other grades of flash welded chain (Grades 2, 3, and Oil Rig) are commercially available.

Dialock chain is manufactured using a series of heavy drop-

hammers by taking a section of bar stock, drop-forging it, and
heat-treating to form the male member. The female is formed by
forging and piercing. The two half links are then mated. Dialock
chain is much more expensive than flash welded chain. The
highest strength flash weld chain is the Oil Rig Quality. This
chain is covered by an API specification which has been approved. High strengths are currently under field test.

Chain is subject to a number of tests before delivery. Proof load
is a load slightly higher than the yield point of the material,
whereas break test is the advertised failure point. The actual
failure tension is approximately 7% higher. Loose studs, and
especially missing studs, greatly reduce the chain's life. When
chain is bought, it should be certified by one of the testing
societies, such as the American Bureau of Shipping (A.B.S.). It
should be closely inspected during recovery operations and any
loose or cracked studs should be replaced. Chain wear should be
monitored and chain should be replaced when the diameter has
been reduced to values specified by the testing societies. Quan-
titative values on the fatigue life of chain have not been determined.

Connecting Links and Shackles

There are a number of connecting links and shackles available
for joining two lengths of anchor chains. Although the catalog
proof loads and breaking strengths of these connections are
indicated to be equal to or greater than a comparable size of
chain, the fatigue life characteristics are considerably less. Of all
the types of chain-detachable connection lengths, the Kenter
link has shown to be the best all around (Figure 1-84). Unfortunately, the Kenter link will not fit some wildcats, especially
those built in the United States before the mid and late 1960's.

Any kind of connecting link should be thoroughly inspected,
x-rayed, and magnafluxed. Failures in mooring lines often occur
in connections. When ordering chain mooring line, as few as
possible connecting links should be used. Figure 1-85 shows a
number of shackles and connectors used with chain and wire
rope. Usually a combination of shackles and/or connectors is
needed to attach chain to wire rope and chain or wire rope to

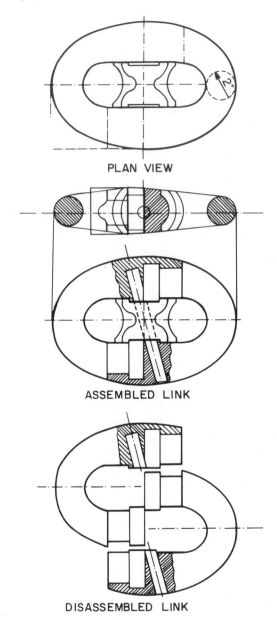

PLAN VIEW

ASSEMBLED LINK

DISASSEMBLED LINK

Fig. 1–84   *Kenter connecting link.*

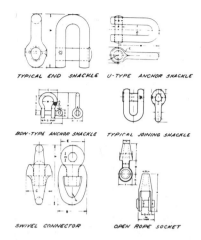

TYPICAL END SHACKLE   U-TYPE ANCHOR SHACKLE

BOW-TYPE ANCHOR SHACKLE   TYPICAL JOINING SHACKLE

SWIVEL CONNECTOR   OPEN ROPE SOCKET

Fig. 1–85 *Mooring line connectors.*

anchors. These connectors are also used to connect pendant lines to the anchor. The importance of connecting links cannot be overstressed.

Design and Analysis of Mooring Systems

In designing and analyzing mooring systems, it is first necessary to determine the environmental force conditions which the vessel and mooring system must withstand. As an example, the North Sea offers some of the most severe environmental conditions the drilling industry has encountered. Table 1-4 is an example of what can be expected for the United Kingdom side of the North Sea. The interrelation between wind velocity and wave height and period is based primarily and approximately on hindcasting techniques and is only applicable to this area.

In analyzing a mooring system, there are two classes of forces which the mooring system must resist. The first is that of "quasi-static" forces such as wind, wave drift, and current. The second consists of the dynamic loads that are induced by vessel surge, sway, roll, pitch, yaw, and heave. The greatest uncertainty in analyzing a mooring system is in calculating the quasi-static and dynamic environmental forces on the rig.

TABLE 1-4

Approximate Wind-Wave Correlation For U.K. North Sea Area (Partially Based On Hindcast)

		Wind			Wave (ft. and sec.)					
		Velocity (Knots)			Significant		Average of Upper 10% Height	Maximum		Seamen's Wind Term (nominal)
Condition	Beaufort Scale (nominal)	10 min. Mean	1 min. Average	Effective	Height	Period		Height	Period	
A	7	31	37	33	17	8.2	22	32	11.2	Moderate Gale
B	8	37	44	40	22	9.3	28	41	12.7	Fresh Gale
C	10	52	62	56	35	11.7	45	65	16.0	Whole Gale
D	12	68	82	74	50	14.0	64	93	19.1	Hurricane

Notes:—Maximum wave height is 1.86 times the significant height. Wave periods are for deep water and 1.99 times the square root of the wave height.

—One minute wind velocity is 1.20 times the 10 minute mean. Effective velocity is 0.6 times 10 minute mean plus 0.4 times 1 minute mean.

—Instantaneous gust of winds may be 40 to 50 percent higher than 10 minute mean.

A number of mooring computer programs are available to the industry which calculate the catenary configuration of the mooring lines and the interaction of the spread mooring pattern, and thus resolve this into restoring force depending upon the direction of displacement off the well. This can be done extremely accurately and has been verified by field tests; however, the estimation of wind, current, and wave drift forces, along with the dynamic effect of the vessel, at present leaves something to be desired.

As an example of the basic mechanics of a mooring analysis, consider Ocean Drilling & Exploration Company's (ODECO) new class of drilling vessel, the "Ocean Voyager." This class of vessel has 12 vertical columns, a nominal width and length of 327 feet, is self-propelled, and displaces 18,600 long tons at a drilling draft of 70 feet. Ten of this class of vessel are in operation.

When analyzing the environmental force conditions on a rig, usually the bow, beam, and maximum force attack angle are analyzed. As an example, Figure 1-86 shows wind, wave drift, and current load on the "Ocean Voyager" class for a beam environmental attack only. The current and wind curves were calculated by the American Bureau of Shipping method. The wave drift force was calculated using a North Sea spectrum. A fully developed North Sea Beaufort 10 spectrum (56 knot mean and 35 foot significant wave height or a corresponding 65 foot maximum height) results in a wave drift force of 75,000 lbs., a wind load of 220,000 lbs., and assuming a current of 1½ knots, an additional 125,000 lbs., for a total of 420,000 lbs., as determined from Figure 1-86. This spectrum is outlined along with three other conditions for a beam attack in Table 1-5.

Figure 1-87 shows horizontal movement (double amplitude) for the "Ocean Voyager" class as determined from model tank test for three directions of movement. The sway oscillation (single amplitude) in Table 1-5 was determined from the beam (sway) data plotted in Figure 1-87. The large environmental forces possible in the North Sea as represented in Tables 1-4 and 1-5, in combination with large horizontal oscillations, are the primary factors which make the North Sea a severe test of any mooring system.

The forces just outlined assume that all three quasi-static

TABLE 1-5

Environmental Forces on ODECO "Ocean Voyager" Class For North Sea In Beam Direction

Condition	Beaufort Wind Scale (nominal)	Wind Effecting Velocity (knots)	Wind Force (kips)	Wave Significant Height (feet)	Wave Significant Period (second)	Single Amp. Sway (feet)	Wave Force (kips)	Current Speed (knots)	Current Force (kips)	Total Quasi-static Force (kips)	Remarks
A	7	33	75	17	8.2	22	40	1.0	55	170	Drilling (nominal)
B	8	40	110	22	9.3	4.3	55	1.0	55	220	Drilling
C	10	56	220	35	11.7	10.0	75	1.5	125	420	Standby
D	12	74	380	50	14.0	16.3	115	2.0	225	720	Shutdown

Note:— Wind force is probably conservative per ABS.
— Current force verified by model tests.

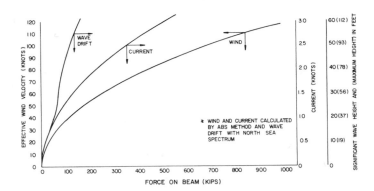

Fig. 1–86 *Wind, wave drift, and current load on beam of ODECO "Ocean Voyager" class*.*

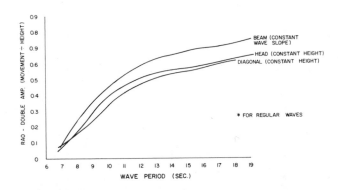

Fig. 1–87 *Response amplitude operator (RAO) vs wave period for surge/sway of "Ocean Voyager" (rough average)*.*

forces (wind, wave drift, and current) are acting on the beam of the vessel. Like all drilling vessels, the "Ocean Voyager" exhibits different loading characteristics from other angles of attack; therefore, Figure 1-86 and Table 1-5 are only valid for a beam environmental attack, which in this rig's case is the worst loading condition.

The next step is to select a pattern and analyze the restoring forces of the mooring systems. The ODECO "Ocean Voyager" class uses an 8 line, 3 inch Oil Rig Quality chain, and symmetrical 30° - 60° pattern measured from the bow, as illustrated in Figure 1-74(f). After the mooring pattern has been selected and the environmental effects have been taken into consideration, the proper pretension in the mooring lines must bedetermined.

Pretension is defined as the initial mooring line tension in all lines with zero offset (no horizontal displacement). By using the one-third break—6% offset criteria outlined in Table 1-3, Figure 1-88 shows pretension vs. water depth for 3 inch Oil Rig Quality chain. The slight curve reversal is due to chain stretch in shallow water. It should be pointed out that pretension based on the latter parameters or any similar parameters varies with water depth, and that the "One-Water-Depth-Theory" is invalid insofar as the pretension-displacements theory is concerned.

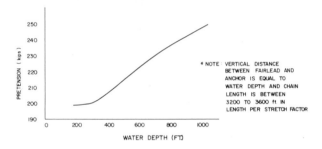

Fig. 1–88 *Required pretension vs water depth for ⅓ rated break strength of 3 in. chain to occur at 6% displacement* (see note).*

The next step is to select a water depth(s) to analyze. Figure 1-89 shows restoring force both with and without the two (2) leeward lines slackened, high line tension (most loaded mooring line), and the high line tension anchor load vs. beam direction horizontal displacement in 600 foot water depths. Once again, Figure 1-89 is only valid for the stated conditions. Note that the proper pretension in 600 feet of water is 222,000 lbs. (Figure 1-88). As stated before, the leeward lines in this condition hinder rather than help the mooring system restore the vessel onto location. The dashed restoring force line assumes that the two leeward lines have been completely manually slackened.

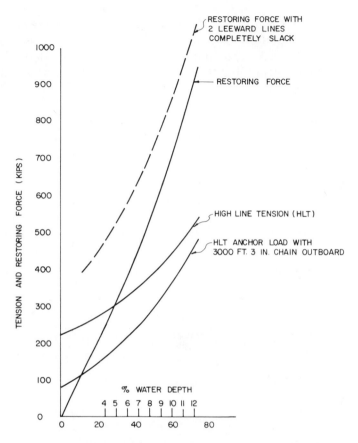

Fig. 1–89   *Tension and restoring force vs beam displacement for the "Ocean Voyager," 30°-60° pattern and 3 in. chain in 600 ft. water depth.*

As can be seen, at small displacements this method of "Dynamic Mooring Assist" significantly helps the restoring force curve; however, at large displacements in which the leeward lines have already been slackened due to displacement, there is little help. In general, it is impossible to completely slacken two leeward mooring lines; however, the dash lines in

Figure 1-89 give some guidance as to the effect of manipulating and reducing high mooring line loads.

By manipulating Table 1-5 and Figure 1-89 together, the mooring system's performance can be analyzed in detail. For example, the Beaufort 10 condition of 420,000 lbs. of "quasi-static" force results in a mean or average displacement of 39 feet (6.5%) and a high line tension of 340,000 lbs. with no leeward line slackening. With the two leeward lines completely slackened, the high line tension is reduced at the mean condition to 260,000 lbs. and 16 feet of displacement (2.7%). This is only the mean condition. With the ± 10 feet of sway about the mean position, the high line tension without slack now varies between 385,000 lbs. (8.2%) and 300,000 lbs. (4.8%).

This 85,000 lbs. of variation about the 340,000 lb. mean is the reason the mooring line's life is relatively short in the North Sea. By slackening the leeward lines, the peak to minimum load is reduced to 290,000 and 235,000 lbs., respectively, with a spread of 55,000 lbs., which is an improvement. The high line anchor load for the two leeward line slackened condition is 95,000 lbs. (minimum) and 170,000 lbs. (maximum).

The importance of the proper pretension cannot be overstressed. In the case just discussed, 222,000 lbs. of pretension resulted in a proper restoring force curve as defined. If the pretension had been higher, the restoring force curve would have steepened as well as the high line tension curve. This would have resulted in excessive chain tensions due to vessel motions at small offsets. Since the mooring system has a negligible to nominal effect on vessel motion, a 20 foot surge at too high a pretension, for example, would result in very large tension fluctuation—and possible failure in mooring line.

On the other hand, if the pretension had been lower, the mooring system would have been "too loose" resulting in too large excursions off the well before sufficient restoring force was supplied by the mooring system. In this case drilling operations would have had to be discontinued due to excessive displacement off the well. In the case of too high pretension, drilling would have had to be discontinued due to overstressing the mooring system.

Vessel Machinery

The use of chain requires a windlass, chain stopper, fair-leader, and possibly a chock or bolster, as shown in Figures 1-90A, B, and C. The chain windlass is used to control movement of the chain during deployment or retrieval. The windlass consists of a power drive (electric or independent motor) and a sprocket (wildcat) with whelps designed for only one size of chain. Chain stoppers, which are a chain locking device, are used neither to remove the load from the wildcat or lock the wildcat. Windlasses do not store chain whereas wire rope winches do.

Figure 1-91 shows a large mooring winch. Winches are usually driven by electric motors and have locking paws that lock the drum rather than hold the wire rope. Electric motors are preferred over independent combustion motors. Hydraulic power has also recently been used.

Fairleads are used to change direction of mooring line movement. They can be of a swivel type as shown in Figure 1-90B, which is the most common, or a nonswivel type which changes

Fig. 1–90A *Releasable chain stoppers (vertical members in front of winlass).*

Fig. 1–90B   *Swivel type fairlead with small portion of anchor bolsters showing.*

Fig. 1–90C   *Double winlass (electrical driven) on semisubmersibles.*

Fig. 1–91 *Large 3 in. diameter and 600 ft. length) wire rope mooring winch. (Courtesy of Skegit Corp.)*

movement in one direction only. Chocks or bolsters can be friction type fairleaders; however, they are not recommended unless absolutely necessary.

Tension meters are desirable to measure chain tension. They can be of a strain gauge, angle inclinometer, or hydraulic load cell type. The measurement point is usually made at the wildcat or winch and does not take into account the friction developed through fairleads, bolsters, or chocks. Until recently, accurate and durable chain tensioning measuring devices were not readily available. Figure 1-92A shows a load cell type of chain tension measurement. Wireline tensioning measuring devices usually operate off a line deflection-load device as shown in Figure 1-92B.

Inspection and maintenance of all the mooring hardware are extremely important since it is exposed to a severe environment and stands idle most of the time. If a good inspection and maintenance program is not carried out, considerable rig time and loss of money can be expected in addition to possible danger to life.

Fig. 1–92A *Hydraulic load cell used for tension read-out on chain windlass.*

Fig. 1–92B *Wire rope tension and read-out device (note wire rope is slack).*

Mooring Line and Anchor Deployment

Almost all floating drilling vessels are unable to place all of the anchors required in a spread mooring pattern. Therefore, anchor handling and tow boats are used for deployment of the system. Anchor handling boats, especially when handling chain in rough environments, should be thoroughly analyzed so that the proper size boat is used.

Figure 1-93 shows the geometry for deploying chain from a semisubmersible drilling rig. The chain is played out of the chain locker, over the wildcat, through the lower fairleader (point A), and down to the ocean bottom at point B. During deployment the chain should not be allowed to "pile-up" at point B or a large bundle or knot will develop. The chain is dragged along the ocean bottom from B to C. Tests from a number of sources have shown that the static coefficient of friction of chain on mud, clay, and sand is approximately 0.8 to 1.0 while the dynamic coefficient of friction varies from 0.13 to 0.74 with a gross average of approximately 0.52. From C to D a catenary is developed by the horizontal pulling force of the anchor handling boat.

Since frictional drag of the chain greatly hinders the deployment process, the anchor should be held as close to the water surface as possible to minimize the required shaft horsepower (SHP) aboard the handling boat.

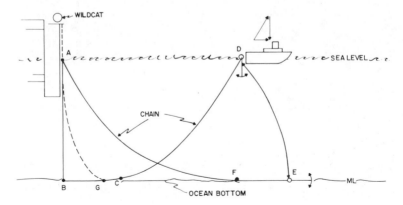

Fig. 1–93 *Chain development geometry.*

Once all of the desired amount of chain has been deployed, the anchor is then swung in an arc from D to E and is initially set by the anchor handling boat. Pendant wire attached to the anchor is used to lower the anchor to the ocean bottom. Proper setting of anchors should be done by deploying the anchor as gently as possible to the ocean bottom rather than by "spudding," since spudding will usually cause anchor damage, fouled mooring lines, and/or damaged pendant wire.

Figure 1-94 shows the proper method for setting anchors. During this process, tow boats are usually attached to the semisubmersible which is at transient draft until all mooring lines are deployed. Upon deployment, the rig is ballasted to

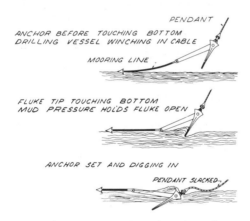

Fig. 1–94   *Sequence for proper setting of anchors from anchor setting boats.*

drilling draft, the anchors are preloaded, and then all lines are pretensioned as the catenary is developed from A to F (Figure 1-93). If the boat has sufficient SHP, a catenary can be developed from A to G; however, for design purposes it is advisable to use the geometry from A to B to C to D, unless the use of a large anchor handling boat can be reasonably assured.

Figure 1-95 is a graphical determination of the required amount of bollard pull necessary to deploy 3 inch Oil Rig Quality chain in water depths of 200 to 800 feet. The solid lines in the

right hand quadrant assume a static condition with a coefficient
of friction of chain on bottom of 1.0 (conservative). The dashed
lines in the right hand quadrant assume that the chain is kept
moving by the anchor handling boat and windlass operator so
that the dynamic coefficient of friction of chain on bottom is
0.52.

The left hand quadrant in Figure 1-95 is a variable scale for
bollard pull per SHP of the anchor handling boat. It is important
to delineate between total installed horsepower and that horsepower which is generated to the propeller shaft(s). The reason
for the large range of bollard pulls per SHP from 12 to 32 is
created by the effect of various propeller types and sizes, use or
nonuse of kort nozzles, and the shape of the hull. High speed,
small propellers without the aid of kort nozzles or a hydroconic
hull will result in low bollard pulls. The use of kort nozzles,
controllable pitched propellers, and a hydroconic hull all contribute to higher bollard pulls per SHP.

Most boats built during the 1960's will exhibit bollard pulls in
the range of approximately 18 to 25 lbs. per SHP in calm water.
Some of the new vessels built for the North Sea will exhibit the
higher bollard pulls per SHP. In heave seas, especially on the
beam or the quarter beam, effective bollard pull is *significantly*

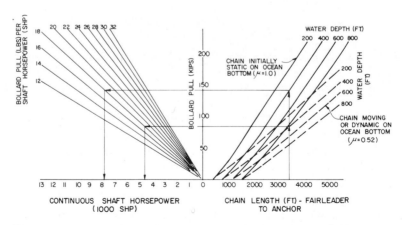

Fig. 1–95   *Anchor handling boat shaft horsepower requirements to deploy 3 in. chain. per Figure 1-93 geometry.*

reduced. This should be taken into account since the anchor handling boat will be deploying anchors in all directions, and therefore will experience beam and quarter beam seas. As a general rule, bollard pull is reduced by approximately 15 to 25% in a beam or quarter beam sea. It is recommended that the anchor handling boat have a bow thruster for better directional control.

As an example, deployment of 3,500 feet of 3 inch chain in 400 feet of water will require a bollard pull of 142,000 lbs. in a static condition or 85,000 lbs. if the chain is kept moving on the ocean bottom. This emphasizes the need for close coordination between the crew and skipper aboard the anchor handling boat and the windlass operator aboard the drilling vessel. If the latter amount of chain is deployed by a conventional anchor handling boat, a minimum of 8,000 SHP would be advisable. If a 5,000 SHP boat is used, then the chain must be kept moving so that the friction on the bottom of the ocean is maintained at the reduced level. In the latter case, if the boat should stall or change heading, it may be necessary that the drilling vessel haul in the chain to the point where the boat can thus pull out the chain again, and keep it moving to full deployment.

Once the chain has been deployed and the anchor initially set, the anchor should be preloaded to its maximum expected load as may be determined from calculations previously discussed. Once the anchor has been preloaded, then the required pretension should be pulled and held. If the chain is deployed such as shown in Figure 1-93, a certain amount of chain will be brought inboard upon pretensioning. Figure 1-96 shows the amount of 3 inch chain that can be expected to be brought inboard subject to the geometries shown in Figure 1-93 from A, B, C, and D to A, F, and E.

Also shown in Figure 1-96 are nominal pretension requirements based on a policy of maximum chain "working" load (350 kips) at 6% horizontal offset as shown in Figure 1-88. For example, in 400 foot water depths, a pretension of 207,000 lbs. will require bringing 300 feet of chain inboard. The amount of chain that is brought inboard can be reduced, especially in deep water, if the anchor handling boat can generate a catenary on the rig end as shown in Figure 1-93, from A to G. Therefore, Figure 1-96 is very conservative; however, to generate any more than a

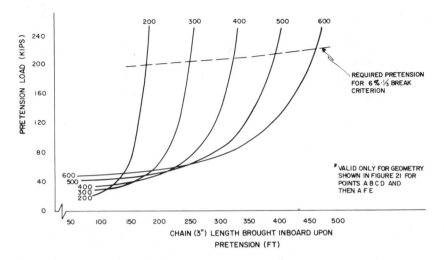

Fig. 1–96   *Determination of amount of 3 in. chain brought inboard upon pretensioning after deployment*.*

nominal size catenary at the rig end will require large bollard pulls which may be uneconomic insofar as boat sizing is concerned. The aid of a tug on the anchor handling boat may be desirable depending on the economics of the situation. Also, calculations may show that not all the chain onboard the vessel need be deployed.

5

Dynamic Positioning

T. R. Reinhart
The Baylor Company

Dynamic positioning, the concept of remaining in one spot at sea without anchors, was originally proposed for the National Science Foundation Project Mohole to drill through the earth's thinner mantle underlying ultra deep 20,000 foot water depths. Although Mohole was canceled, dynamic positioning was used for several small core boats because it provided deep water mobility. The use of dynamic positioning on the Glomar Challenger in deep ocean coring helped bring discoveries adding to man's knowledge of the world.

Today, dynamic positioning is being applied to the deep water search for natural resources. New, scientific disciplines are developed for position determination and position keeping. Large thrusters have been designed. Power plants and total systems with 10,000, 20,000 or perhaps even 40,000 horsepower are instrumental in position keeping for indeterminate periods of time, even in areas of significant weather activity.

BACKGROUND AND HISTORY

Dynamic positioning is a technique of automatically maintaining the position of a floating, unanchored vessel within a specified tolerance by using generated thrust vectors to counter the forces of wind, wave, and current forces tending to move the vessel away from the desired location. Design evolution and improvements in reliability allow station keeping by dynamic positioning for extended periods. Increases in available power and advances in sophistication of control equipment allow station keeping to be maintained in increasingly severe levels of wind and wave intensity.

Position is usually defined in terms of percent of water depth. Percent of water depth is the horizontal positional error divided by the water depth multiplied by 100. Position error, expressed in percent of water depth, is preferred because it defines the position and it is also related to the stress level in the riser or drill pipe. Generally, accuracies of the positioning system itself are about 1 percent and positioning to an accuracy of 1 percent or less is possible only in calm water and wind.

Five percent represents a common maximum permissible error with respect to permissible stress levels in the tubulars running from the ship to the ocean floor. At 5 percent of water depth, the angle off verticle of the drill pipe would be 3 degrees, a very small angle. At 10 percent of water depth, most drill strings would become bent or damaged. With significant wave action and vessel motion due to wave, the string may be lost.

Increasing water depth makes the task of dynamic positioning much easier because the same percent of water depth permits more movement in deeper water. For instance, given a 5 percent accuracy requirement a nearly impossible accuracy requirement of 5 feet is set up for 100 feet of water. Similarly, with the same 5 percent requirement applied to 1,000 feet of water an off-the-hole allowance of 50 feet is given, a much more reasonable tolerance. For 10,000 feet of water, the allowable radius of surface movement is a generous 500 feet.

This desirable feature of deeper water, however, is partially offset by some of the difficulties found in position determination in deep water.

Mohole Tests

The first dynamic positioning involved a feasibility study and test for the now defunct Project Mohole in March and April, 1961. In tests, the barge Cuss I was maintained on location by 4 harbormaster units controlled essentially by manual control from visual interpretation of position reference displays. The 4 harbormaster units were attached to the 4 corners of the vessel. Position was indicated by radar reflecting surface buoys and sonar signals from a ring of subsurface buoys anchored at a relatively shallow depth in the very deep 12,000 feet of water off La Jolla, California and Guadeloupe Island, Mexico.

The Mohole tests were successful in that they did show that dynamic positioning could be effective in recovering drilled cores from such deep water depths. Difficulties encountered in both position determination and automatic operation of the positioning equipment (thrusters) pointed out the need and direction for further improvement in these areas. The great water depth of the tests, 12,000 feet, in most respects was an advantage. It allowed an operating radius of 600 feet at the 5 percent of water depth limit.

Core Boat Eureka

Figure 1-97 shows the first fully successful dynamically positioned vessel, the M/V Eureka, developed by Shell Oil Company. This small 36 ft. x 136 ft. core boat was originally designed to be manually operated, but early demonstration of the difficulty of manually coordinating the two 200 horsepower steerable thrusters resulted in a fully automatic, nonredundant, closed loop control system. The Eureka first began operating in May 1961. Separate analog controllers were used for each of the three degrees of freedom of motion—surge, sway, and yaw. Standard startup procedures were used for optimizing rate, proportional band, and reset settings for each of the three controllers.

The Eureka was a successful vessel, having drilled core holes in as little as 30 feet and as deep as 4,500 feet of water. Drilling has been performed with winds up to 40 mph and 20 foot swells. As many as 14 core holes have been drilled in one day.

Caldrill I

The Caldrill I was the second dynamically positioned vessel. Figures 1-98 and 1-99 show the vessel at dock and also drilling while dynamically positioned. The Caldrill I, 176 ft. long and 33 ft. in beam, was powered by four fully steerable 300 horsepower

Fig. 1–97   *Core boat Eureka.*

Fig. 1–98   *Drilling vessel Caldrill I at dock. (Courtesy of Caldrill)*

Fig. 1–99   *Caldrill I drilling with dynamic positioning. (Courtesy of Caldrill)*

thrusters. The vessel was equipped to drill 6,000 feet with 4½ inch drill pipe. The Caldrill I has operated in 25 foot swells off Nova Scotia and has maintained position in the face of wind gusts up to 60 mph and ocean currents of 3 knots. The vessel is still in service and has operated in the oceans of the world.

For reliability for live well operations, the Caldrill I has four 300 horsepower thrusters with two separate dynamic positioning systems. The thrusters on both the Caldrill I and the Eureka are open propeller type and retracting. Two taut line systems on the Caldrill I provide position sensing.

Other Ships

The French vessel La Terebel made its debut in December 1964. This vessel was 85 meters long, 12 meters in beam (278 ft. by 39 ft.) with two 300 horsepower azimuthing thrusters. The thrusters were directly driven by diesel engines.

In 1967, the Mission Capistrano, under U.S. Navy contract to Hudson Laboratory, was used for anti-submarine warfare work. The Mission Capistrano had two 1,000 horsepower azimuthing thrusters. The position computation system was made by General Motors AC Division.

Challenger

In August 1968 the Glomar Challenger appeared, 400 ft. long by 65 ft. beam, displacing 10,500 long tons (Figure 1-100). The Challenger had two main screws with 2,250 horsepower each and was powered by three GE 752, 750 horsepower DC traction motors. Four tunnel thrusters were each powered by 750 horsepower traction motors to provide 3,000 horsepower for thwartships thrust. The Challenger has used two acoustic positioning systems—a phase comparison system, and a time of arrival system. Position computation was provided by a General Motors AC division system.

Saipem II, Pelican, Sedco 445

In 1971 and 1972, three major drillships having an obviously similar mission but showing distinct differences in design

Fig. 1–100   *Glomar Challenger. (Courtesy of Globel Marine)*

philosophy were brought into existence. These ships were the
Sedco 445, the Pelican, and the Saipem II dynamically
positioned ships (Figures 1-101, 1-102, and 1-103). It is interesting to note in Figure 1-104 that each ship has about 14,000 tons
displacement, each is between 400 and 500 ft. long, and each
ship is about 70 ft. beam. Cruising speed on all three vessels is
about 14 knots. However, the similarities end here. Saipem II
uses 4 Voith-Schneider omni-directional thrusters, the Pelican
uses five 1,500 horsepower fixed tunnel thrusters and 2 main
screws all with controllable pitch (CP), and the Sedco 445 uses
nine 800 horsepower fixed thrusters with kort nozzles and a
reversing, variable rpm fixed propeller drive on the thrusters
and 2 main screws.

Specific references to design characteristics and operating
experience of these vessels may be found in OTC (Offshore
Technology Conference) papers beginning with 1970.

Fig. 1–101   *Dynamically positioned Sedco 445.*

ELEMENTS OF DYNAMIC POSITIONING

The three elements basic to dynamic positioning are:
1) Position measurement or the continuous measurement of position with respect to ocean bottom borehole.
2) Control response or determination of correct thruster response from position measurement.
3) Thrust response or realizing body forces by expenditure of energy as a result of control command.

Taut Line System for Position Measurement

The Taut Line System uses a small diameter line attached to a weight on bottom. The weight is of such a size that its dead weight will stress the line 10 to 20 percent of its breaking strength. Similarly, on top a counter weight is rigged to effect a

Fig. 1–102   *The Pelican.*

constant tension of 5 to 19 percent of the breaking strength of the line. For example, a counter weight of 500 lbs. might be used to tension a ⅛ inch line against a 1,000 lb. weight on bottom. Under this load of tension, the line at the surface points to the weight on bottom without moving the weight.

Table 1-6 gives some values for the weight on bottom for various wire line sizes. This weight ranges from an 800 lb. weight for a 400 lb. line tension for a ⅛ inch line to 25,000 lbs. for a 7 inch line. Iron or steel items make a good weight, losing only 12 percent of their weight due to buoyancy. In Figure 1-105, the counterbalance weight is on a multipart line of 8 parts. The 8 parts increase the necessary weight 8-fold to a value of 1,600 lbs. This 1,600 lbs. is provided by lead shot filling the weight assembly.

Fig. 1–103   *Drillship Saipem Due.*

A two-axis inclinometer is fitted with a guide and gimbaled to pick off the angle of the wire line with respect to the vertical. The vertical angle is referenced to a shipboard coordinate system of bow-stern axis inclination and post-starboard axis inclination. A knowledge of the water depth and a little geometry allows calculation of the position of the vessel with respect to the weight on bottom (Figure 1-106).

Simple zero shift of the two axes to compensate for the initial offset between the weight and the rotary table makes the taut line system read out the position of the rotary table with respect to the hole location on bottom. Dual units as shown in Figure 1-107 may be used for reliability and to allow large heading changes. A

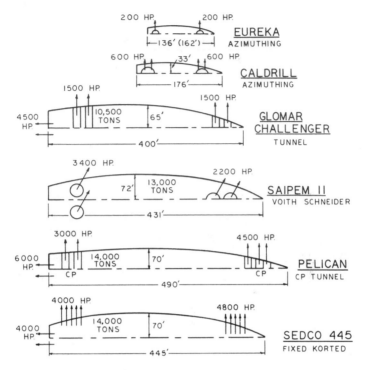

Fig. 1–104   *Vital statistics of operational dynamic positioned ships.*

partial listing of the advantages and disadvantages of the taut
line system follow.

Advantages of taut line:
a) Simple and economical
b) Visible evidence of operation
c) All sophisticated equipment on deck
d) Rapid deployment
e) Works well in shallow water
f) Immune from underwater noise

Disadvantages of taut line:
a) Mechanical taut line or tensioning system can fail
b) Large heading change must be made by stepping two units
c) Susceptible to error due to current drag

TABLE 1-6

Sample Tension Vs. Wire Line Size For Taut Line Unit

Wire Size	Wire Construction	Area	Breaking Strength	Tension for 32,600 psi Stress	Safety Factor
⅛	1 x .19	.012272	2,100#	400#	5.0
¼	1 x 19	.049087	8,400#	1,600#	5.0
⅜	6 x 19 XIPS	.110447	15,100#	3,600#	4.1
½	6 x 19 XIPS	.196350	26,600#	6,400#	4.1
⅝	6 x 19 XIPS	.3068	41,200#	10,000#	4.1
¾	6 x 19 XIPS	.4418	58,800#	14,400#	4.1
⅞	6 x 19 XIPS	.601	79,600#	19,600#	4.1
1	6 x 19 XIPS	.785	103,400#	25,600#	4.1

The taut line inclinometer, normally a gravity sensing device, cannot distinguish between a true angle from the verticle and a horizontal acceleration. Thus, the effect of the horizontal accelerations accompanying a peak-to-peak surge or sway of 5 feet and a period of 12 seconds produces an oscillating apparent sway of 1.9 percent of water depth. This apparent sway subtracts from the 5 foot peak-to-peak true motion.

. The 1.9 percent of water depth error is read out as feet according to the water depth. At 180 foot water depths in the above example, actual surge motion is exactly cancelled out by the surge acceleration induced effects in the inclinometer. At a greater depth than 180 feet, the acceleration induced error exceeds the actual surge motions. The problem is solved by two alternative methods: using a gyroscopically stabilized inclinometer which is insensitive to horizontal acceleration, or by simply limiting the response of the vessel only to inputs of longer periods than wave motions.

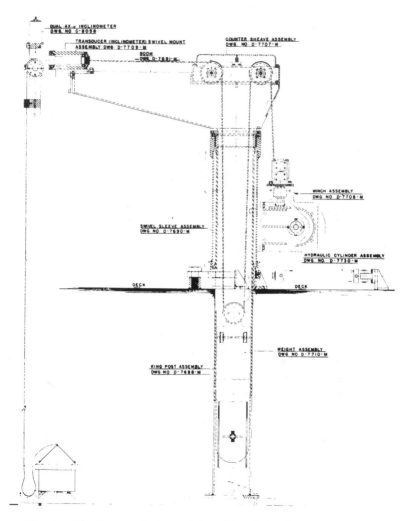

Fig. 1–105   *Counter-balanced Taut Line Unit. (Courtesy of Baylor)*

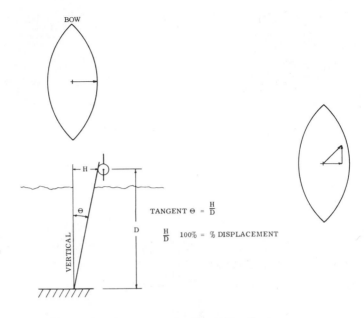

Fig. 1–106   *Geometry of a Taut Wire System.*

Acoustic Position Reference Systems

While the taut line system requires use of a wire to mechanically communicate position from the ocean bottom to the surface, the acoustic systems use underwater sound waves as the required communication link. Acoustic waves travel considerably faster in water than in air. The velocity for sound in water is almost 3,300 mph, about 4.4 times faster than the speed of sound in air. The short base line acoustic position reference system illustrated in Figure 1-108 has an acoustic beacon located on the bottom and a vessel with four hydrophones on the sea surface.

The beacon shown in Figure 1-109 transmits acoustic pulses at regular intervals to the hydrophone array. When the vessel is directly over the beacon (or drill hole) acoustic signals arrive at all hydrophones simultaneously. When the vessel is displaced away from the hole, the nearest hydrophone receives the acous-

Fig. 1–107   *Dual Taut Line installation. (Courtesy of Baylor)*

tic wave front first and the furthermost hydrophone receives the acoustic wave front last. These differences in time, from the nearest hydrophone to the most remote hydrophone, are operated on by a signal processor whose output is directly proportional to the percent water depth positioning error. Two sets of hydrophones at right angles to one another give the position error in two coordinates, X and Y, corresponding to bow-stern error and port-starboard error. Position is displayed on the control indicator shown in Figure 1-110.

If a vessel is considered with its hydrophone array directly centered over the beacon, but with a slight roll, it is easily seen that a roll to starboard will give the starboard hydrophone a deeper submergence in the water while, at the same time, the port hydrophone is submerged to a shallower depth. If the starboard hydrophone is deeper then it is also closer to the beacon and the port hydrophone is further away from the beacon.

This change in distance of the hydrophones from the beacon

RS-5 EQUIPMENT

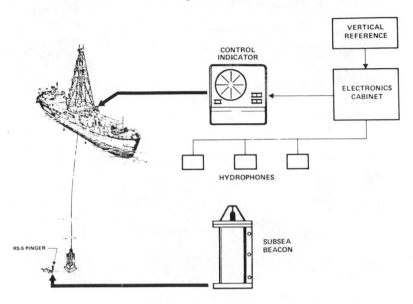

Fig. 1–108   *Short baseline acoustic system. (Courtesy of Honeywell Inc.)*

due to roll makes the short base line acoustic system require a
Vertical Reference Unit (VRU) to measure pitch and roll. This
allows the signal processor to eliminate the pitch and roll induced errors in the hydrophone signals.

Since the acoustic beacon may not be installed over the bore
hole or the hydrophone array may not have been centered under
the rotary table, a means of dialing in offsets in X and Y in feet is
provided. Up to 100 feet of offset compensation is usually allowed. The sum of the beacon offset and the rotary table offset
also should be less than 20 percent of the vertical separation
between the beacon and the hydrophones. On a large, deep
submergence semisubmersible rig with a tall B.O.P. stack assembly, the distance between the hydrophones and the beacon
may easily be 90 feet less than the water depth. This 90 feet may
result in a low separation for the hydrophone to beacon distance,
even though actual water depth would not lead one to suspect
such a problem.

Fig. 1–109   *Acoustic beacon. (Courtesy of Honeywell)*

The ususal pitch and roll reference used is a gravity sensitive unit. This unit will respond to surge and sway accelerations to produce an acceleration error identical to one developed in a gravity sensitive inclinometer of a taut line system. As with the taut line system, the two means of dealing with the problem are to eliminate all "short" period inputs by filtering to respond only to very low frequency inputs or to use a gyroscopic vertical reference unit (VRU) which is insensitive to horizontal accelerations.

This gyroscopic vertical reference unit is quite costly and may have a life of only 20 to 50 days due to bearing life in the gyroscope. A working arrangement has been to use a conven-

Fig. 1–110   *Acoustic system control indicator. (Courtesy of Honeywell)*

tional pendulous VRU during normal weather when motions are
low. The gyroscopic VRU is then turned on to operate only
during bad weather. The life of the gyroscopic VRU is conserved
by using it only whenever optimum positioning performance is
required.

A good, short base line acoustic positioning system can maintain a position sensing accuracy of 1 percent of water depth (1%
WD) under favorable conditions. This accuracy relies on a correct hydrophone and VRU installation. Accuracy requirements
on installation are in the order of magnitude of 1 degree of
angular measurement and distance accurate to 0.2 percent (1
inch in 40 feet). The most stringent requirement on installation
is mounting the VRU parallel to the plane of the hydrophones.
The total cumulative allowable misalignment (electrical and
mechanical) for the VRU reference plane to hydrophone plane

alignment is ± 0.1 degree for a typical system. Achieving this 1/10 of a degree alignment is simple on land, but will call for some thought and care if the vessel is afloat at the time of hydrophone installation.

In the operation of an acoustic system, many mathematical procedures are performed. For pitch and roll compensation, the output of the VRU is subtracted from the processed hydrophone signals. One degree of pitch or roll acts exactly as a 1.75 percent of water depth (1 degree) change of vessel position if not compensated. Four degrees of pitch or roll would be read out as 6.98 percent of vessel position error if not properly compensated. To ease the calculation involved in determining the vessel position from the processed hydrophone signals summed up with the VRU pitch and roll signals, a very excellent small-angle approximation is made that the sine of the angle, the tangent of the angle, and the angle itself are all equal. As shown in Table 1-7, 6 degrees corresponds to a vessel position of 10.47 percent of water depth, a point at which the sine approximation is 0.18 percent low and the tangent approximation is 0.36 percent high.

The small-angle approximation for the sine shows a 1 percent accuracy up to vessel position errors of 17.5 percent which is more than adequate for any normal drilling task.

TABLE 1-7

Sin, Tan, and Θ for Small Angles
Showing Closeness of Simple Approximation

Θ (Degrees)	0	1	2	4	6	10	20
Θ (Radians)	0	0.0174	0.0349	0.0698	0.1047	0.1745	0.3490
Θ (% WD)	0	1.74	3.49	6.98	10.47	17.45	34.90
Sin Θ (% WD)	0	1.745	3.490	6.976	10.453	17.365	34.202
Tan Θ (%WD)	0	1.746	3.492	6.993	10.510	17.633	36.397

Example: At 6°
     6° = 0.1047 Radians x 100 = 10.47
     Sin 6° x 100   = 10.45
     Tan 6° x 100   = 10.51

The offsets of the center of the hydrophone array to the rotary table and that of the beacon to the subsea well bore must be subtracted from the measured position to obtain the true vessel position. As mentioned previously, the sum of these two offsets expressed as a percent of *the vertical separation between the hydrophones and the beacon* may be limited to 20 percent or less. In the cases where an acoustic system is to be used with an anchored semisubmersible, the effect of hydrophone and beacon offset and shallow water should be investigated by the user and the equipment supplier. This investigation is necessary to check the following items:

1) Reduced output of beacon at angles from the beacon axis of 20 percent or greater.
2) Reduced sensitivity of hydrophones at high angles of incidence of 20 degrees or greater.
3) Departures from linearity of sine, tangent, and angle approximations at 20 percent of water depth or greater.
4) Possible multiple path reflections such as beacon to vessel hull reflected back to sea bottom and then reflected up to hydrophones.

Note: The acoustic system may have a 100 percent of water depth acquire mode that operates as an exception to the above comments. It does so by sacrificing accuracy, which is not necessary in the acquire mode.

If a vessel is operated with the percentage ratio of the combined hydrophone to rotary table offset and bore hole to beacon offset exceeding 20 percent, the vertical motion of the vessel due to heave or tides which makes an effective change in water depth may yield a source of error. The acoustic system with a large percent total offset will fail to subtract the correct percent offset for the variation in water depth. The error is correctable only by feeding the true water depth, including tide and heave, into the system as a variable in the determination of the percent offset.

The offset referred to is the large hydrophone and beacon offset. This water depth input, normally a constant set into the system, would have to be an onsite, real-time measurement such as a bottom echoing fathometer might provide. To the author's knowledge, at least one instance is known where this shallow water complication was an insurmountable obstacle to the desired system accuracy. The amount of error produced is closely approximated by the equation:

$$\text{WD Error} = (\% \text{ Effective Offset}) \times \frac{\text{Heave} + \text{Tide Motion (Ft.)}}{[(\text{Effective WD (Ft.)}]}$$

where the % Effective Offset is the combined beacon and hydrophone offsets as a percent of water depth.

For example, consider a vessel with a 20 percent effective offset in an area with a 10 foot tidal change and a 75 foot effective water depth. The tide apparent position error will be:

$$\% \text{ WD Tide Error} = (20\% \text{ Offset}) \times (10 \text{ ft.}/75 \text{ ft.}) = 2.7\%$$

This tide or heave error is not too severe if the percent offset is kept below 20 percent.

The spectrum of acoustic frequencies usable for the short base line systems probably lies between 10 kilohertz and 100 kilohertz. The choice of frequencies in this region is not as simple as it first seems because one set of factors is optimum at the high frequencies while another set of factors is optimum at low frequencies. In the matter of wave length, for instance, at 10 kilohertz the wave length in water is about 6 inches, while at 100 kilohertz the wave length is 0.6 inches. The hydrophones should have a very low sensitivity from ship or surface noise and have a high sensitivity directed downward in a 10 to 30 degree cone.

Since the hydrophone is analogous to an antenna, the higher frequencies result in smaller hydrophones and more easily obtainable directional characteristics. Thus, physical hydrophone design favors the higher frequencies because of the shorter wave length. This same argument holds true for the bottom located beacon and its directional characteristics.

Noise levels also favor the use of high frequencies as the rig generated noise is predominantly low frequency and the amount of interfering vessel generated noise is significantly reduced at the higher frequency. This question of noise requires further investigation to define the magnitude, source, and frequency of the noise such as from the following sources:

   a) Propeller cavitation,
   b) Thruster final drive gears,
   c) Final drive chain reductions,
   d) Other machinery noises.
   e) Gas bubbles.

The attenuation characteristics of water favor the use of low frequencies because as the frequency is increased the transmission losses increase strongly. It is seen that the choice of hydrophone frequency is not a simple one due to the absence of a single optimum condition. Generally speaking a less expensive, more precise high frequency system can be made for shallow to moderate water depths. For deeper systems, a more expensive low frequency system is indicated to take advantage of the lower transmission losses at the lower frequency.

Other Positioning Systems

Long base line acoustic systems are similar to short base line systems turned inside out or upside down. For example, consider a long base line system in which a single beacon is suspended underneath the ship and the ship is centered within a one mile square pattern of anchored hydrophones. The hydrophone signal is connected by cable to a surface buoy and radio transmitter which transmits back to the ship. Every second a short acoustic pulse is transmitted from the beacon. At the time the acoustic pulse reaches the hydrophone, the radio transmitter relays its arrival back to the ship.

It is easy to see that the delay time from the start of the beacon transmission to the receipt of the signal transmitted from the hydrophone is a direct measurement of the distance to the remote hydrophone. In fact, four radio channels effectively allow measurement of the distance to each hydrophone, assuming a constant speed of sound in water.

The above described system is an upside down version of the short base line system. This system seems to have great accuracy since it measures the distance to four remote subsea markers. It does not contain any pitch and roll inaccuracies except for the actual X and Y motion of the beacon as it moves in arcs formed by the pitch and roll produced by the radius of the beacon to the center of pitch and roll. The great disadvantage of this system is the need to deploy the elements of the long base line—the subsea hydrophones with their mechanical and electrical connection to the buoys and the general complexity of the active elements of the remote transmitters.

Emplacement of a long base line system is several orders of magnitude more difficult than emplacing the single beacon used in the short base line system. The short base line is "short" out of necessity because its base line is on the vessel.

Radar

Radar is an obvious possibility as a positioning system, but it suffers the common shortcoming inherent of all long base line systems. If some natural target does not exist, then a system such as buoys with radar reflectors must be emplaced and the system becomes dependent on the buoys staying in place through wind, wave, and storm. Radar vessel positioning, however, can work if there are dependable base line targets.

Decca, Raydist, or Loran

The Decca, Raydist, and Loran systems are all long base line systems which rely on two or more transmitters separated by a number of miles. These systems all have the capability to provide for dynamic positioning if the vessel is within the operating range of required accuracy. There is some concern over interruption of broadcasting and the dynamic positioning dependence on the remote transmitter. The patterns established by the multiple transmitters may not have constant accuracy over all portions of the pattern, and the location of use must be within the more accurate areas of the system pattern. The area of use may also be within the range of existing navigational systems (such as the North Sea). For instance, a successful dynamic positioning test using one of these systems has been conducted in the Rhine River.

Inertial Systems

Intertial systems, as developed for submarine navigation, undoubtedly could perform a useful role in dynamic positioning. The two elements of cost and restricted information, however, have made the inertial system unavailable. Its possible future use coupled with satellite navigation is reviewed in the following satellite navigation section.

Satellite Navigation

Satellite navigation entails communication with a satellite passing overhead and the determination of the latitude and longitude of the vessel at the time of the pass from information transmitted from the satellite to the vessel. Several interesting factors accompany the use of satellite navigation:

a) A satellite must be over the vessel for a position to be given. The satellite may be either an orbiting satellite making an overhead "pass" or it may be a synchronous satellite. The synchronous satellite has a velocity exactly equal to the earth's rotation speed, thus making the synchronous satellite appear motionless in the sky.

b) A single "fix" is accurate only to about 500 ft.

c) The average of a number of "fixes" will be accurate to about 200 ft.

d) A satellite's orbit is about 90 minutes. It may or may not be within range on the next pass.

e) Several satellites are used as navigational satellites. In general, a satellite will be within range to transmit navigational data about 20 times a day.

It is obvious that satellite navigation is much too inaccurate and "fixes" every hour or two are not frequent enough to permit satellite navigation to play a role in dynamic positioning at the present. However, with continued improvements that might be expected, the combination of satellite navigation and inertial systems may complement each other.

For example, if satellite navigation could give a "fix" that is sufficiently accurate, as in very deep water, then a short-term inertial system could hold position until it is updated by another satellite pass. This system could have high reliability and yet have no necessary subsea reference points. Its use would undoubtedly be limited to deep water greater than 1,000 ft. to tolerate the system's large inaccuracy.

VESSEL HEADING REFERENCE

In addition to a position reference, a vessel heading reference is necessary (Figure 1-111). Fortunately, existing standard marine gyrocompass systems fulfill this need very well. A sine-cosine potentiometer is driven by a repeater from the main gyrocompass. The sine-cosine potentionmeter is used instead of a 0-360 degree potentiometer to eliminate the zero to 360 degree full scale jump in output that would occur when the vessel is heading yawed about a true north heading. The gyrocompass can drive other ships' navigation gyro repeaters without any interference from the operation of the dynamic positioning equipment.

Fig. 1–111 *Gyrocompass direction reference. (Courtesy of Sperry)*

DYNAMIC POSITIONING CONTROLLER

Purity of Position Determining Signals

Dynamic positioning is a basic control problem in three degrees of freedom. A free body has six degrees of freedom in motion: surge, sway, heave, pitch, roll, and yaw. The dynamic positioning system will be concerned only with surge (bow-stern translational motion), sway (port-starboard translational motion), and yaw (rotational motion about the vertical axis). No attempt will be made to control pitch, roll, or heave by the dynamic positioning system.

A first requirement of dynamic positioning is that the position signals be "pure" and unaffected by changes on the other axis. In this regard, the sensing systems and their processing should be examined in detail. For example, it was discovered several years after the Eureka became operational that an unanticipated operating mode led to a point of instability due to cross axis modulation of the two translational axes. This instability occurred after the crew had some trouble with the taut line getting tangled up with the drill string. The Eureka's crew found that they could eliminate the problem by dialing in a large starboard offset after setting their taut line weight on bottom. This large offset would cause the vessel to move sideways away from the weight until an offset of about 30 percent water depth had been reached, keeping the taut line well away from the drill string.

With this large offset, plenty of maneuvering room was afforded and it was virtually impossible for the drill string to foul the taut line. Since designers of the vessel had not anticipated such large offsets, the vessel offset was vessel oriented without heading compensation. It first seemed that when operating with such a large offset, X and Y disturbances were handled normally and the positioning system was in good shape. Then, after 100 or more waves had passed, a large wave would swamp the bow and generate a heading error of 10 or 15 degrees.

At this time, the large X offset would generate a spurious Y position error by coupling the offset into the Y channel by the sine of the heading change. The result was that the vessel would go charging ahead to correct the spurious Y signal and when the vessel regained heading it would have to back up to correct the bonafide Y position error it had just created. If the wind and sea conditions were marginal, the excursion often would be enough to twist off the drill pipe or part the taut line. The problem was diagnosed as failure to provide automatic coordinate conversion for the offset input values. The proper coordinate conversion would make the offset referenced to true North instead of to the vessel coordinate system.

There was no error in the offset as long as there was no error in heading. With a heading error, though, the offset would be referenced improperly to the system and sizeable spurious bow-stern errors would be introduced. A similar cross axis modulation in an acoustic system in which vertical motion from tide or heave produced apparent motion was described in the prior section on acoustic position systems.

Linear System Allows Separate Signal Processing

Dynamic positioning fortunately is mainly a linear system allowing the control independence of surge, sway, and yaw. This is to say that separate control channels may exist for each one of the three degrees of freedom and these separate control channels are independent and uncoupled. (Cross coupling will be introduced in the thrusters to be discussed later).

The measurements of surge, sway, and yaw are introduced to three error comparators, one for each degree of freedom. In these error comparators, the current value of surge, sway, or yaw is compared with the set point value for each coordinate axis. The comparators generate error signals which are proportional to the difference between actual position and the set point. This error signal is the input for a process controller which calculates the correct control response to make the error go to zero or stay as close to zero as possible. In the system above, the surge, sway, and yaw signals are completely independent and the process is handled at three separate systems.

Signal Conditioning

Signal conditioning is performed on the error signals for each of the three error signals of surge, sway, and yaw. It has been recognized that power requirements increase by orders of magnitude if attempts are made to resist wave motion. This generally fruitless attempt to resist wave motions can call for peak power beyond the thruster's ability. This excessive demand can saturate the control system causing it to lose control.

Low pass filtering of the error signals of surge, sway, and yaw can reduce the system response to waves to a more acceptable level. These filters are a careful trade-off, as too much filtering will cause excessive loss to the system response time. A fast response time of 5 to 10 seconds is desirable to counter wind gusts, while a much slower response of 30 seconds to 1 minute may be indicated for wave motion filtering.

Other special signal conditioning may be required, but most low pass filtering requirements are more than satisfied by the low pass wave filter. Examples are smoothing the 0.1 second repetition rate acoustic system position determination or removing the residual 400 hertz content of a taut line transducer output.

Active Wind Compensation

The problem of desiring a slow filter for removing wave motion while desiring a quicker filter to improve wind gust recovery is nicely solved by active wind compensation. For active wind compensation, the wind speed and direction relative to the vessel are fed into a "black box" which performs the translation of wind speed and direction to forces on the vessel. The "black box" must internally have model test results with wind fields or some other source of data concerning the wind forces on the vessel.

In any case, the output of the "black box" is force on the vessel. The forces are then added *after* the low pass filtering to their respective channels of surge, sway, or yaw. Since the wind forces are added to the error signals immediately, without delay, they are acted upon by the system controllers to produce immediate counter forces by the vessel thrusters. The active wind compensation translation in the "black box" does not need a high degree of accuracy. Because there may be inaccuracies in the model, it is a good idea to have the "black box" output about 3/4 of the calculated force so that the active wind compensation error is on the low side rather than producing more force than needed.

Simulation tests have shown that active wind compensation is very effective in improving wind gust recovery. Instead of waiting for the wind to move the vessel and then waiting on the low pass wave filter to sense that the vessel is moving, active wind compensation allows immediate input of wind force information to the system controllers.

Basic Controller Action

For vessel control in dynamic positioning, each degree of freedom is controlled separately. Thus, the overall control is divided into three separate channels. Basically these channels consist of a fore-aft position controller, a port-starboard position controller, and a heading position controller. The force or thrust demand output of these three controllers is input to the thruster logic section which determines the actual thruster response to the output commands.

In a given channel of the control system, the system operates as a conventional three mode controller using proportional, integral, and derivative control. The proportional control applies a thrust command output on the axis which is proportional to the amount of error on that axis. It is seen that if an external force is applied on the axis, the vessel must move off location a distance such that the error times the proportional control gain setting equals the external force.

The proportional control is similar to holding the vessel in position with a spring which is slack with zero force at a zero error input. The only way the proportional spring can develop a given external force is by stretching or developing a position error until there is enough stretch or error to counteract the force. As proportional control has a steady state error proportional to the disturbing force, another control action is used which forces the integral of the error signal to be zero. This integral or reset control action integrates the error signal to generate an output signal proportional to the time integral of the error signal. The integral action eliminates any steady state errors inherent in the proportional only control system.

A third control action used to prevent overshoot and to stabilize oscillatory tendencies of the system is derivative or rate action. The rate action increases the system response by the anticipatory action of its response to the time rate of change of the error signal. Each of the three control channels has its own particular constants determining the control values of its proportional setting (called proportional band), integral setting (called reset), and derivative setting (called rate). The final output of the three controllers includes a command signal in the surge axis for longitudinal thrust in that axis, a command signal in the sway axis for lateral thrust in that axis, and a command signal in the yaw axis for a turning moment in that axis. A familiar example of a single degree of freedom controller operating on many ships is that of an automatic pilot for steering control.

THRUSTER LOGIC

In the previous section it was shown how the dynamic positioning controller generated the three positioning commands of surge axis thrust required, sway axis thrust required, and yaw axis moment required. Multiple thruster choices of different types and styles will be provided to satisfy the thrust requirements. Different types and styles from the thruster logic viewpoint include:

a) The Fixed Type is capable of reverse action. Typical of this type are the Glomar Challenger, the Pelican, and the Sedco 445 in which the ship's main screws are used at reduced power level for the fore-aft surge thrust requirement. Lateral thrust for the sway axis is provided by fixed thwartships thrusters. Yaw axis moment is provided by differential operation of the thwartships thrusters to produce a net moment but no net thrust.

b) For the Azimuthing Type, each thruster is capable of azimuthing to the direction of surge and sway axis vector resultant to provide surge and sway thrust. Yaw axis moment is provided by differential vector action of two or more thrusters to produce yaw moment with no net surge or sway force.

c) For Cycloidal Propellers, a special vertical paddle wheel thruster unit with cycloidal pitch variations is provided. Thruster logic is similar to the azimuthing thruster with the difference that the cycloidal propeller thruster may be 10 times as fast as the azimuthing thruster in responding to thrust direction change. Cycloidal propeller thrust also directly accepts surge and sway commands without their being transformed into thrust magnitude and thrust direction as required by the azimuthing thruster.

Degrees of Thruster Freedom

Thruster logic deals with the total task of developing the thrust and moment to satisfy the dynamic position controllers' three output commands of surge thrust, sway thrust, and yaw moment. These three degrees of freedom command require a like number of thruster degrees of freedom to respond. Thrusters are usually applied so that their total degrees of freedom exceed the minimum. The extra degrees of freedom are called redundant degrees of freedom. A single fixed thruster can thrust in only one direction (reverse is just a negative sign) so it can satisfy only one degree of freedom.

A steerable thruster has two degrees of freedom in that it can independently thrust on two axes simultaneously. An example of three fixed thrusters just satisfying the three degrees of freedom of dynamic positioning is that of a twin screw workboat with a bow thruster. In this instance, the operators have learned to move sideways without turning by countering the moment developed by the bow thruster by running one main engine ahead and one astern at the same speed to develop no net fore-aft thrust. Total thruster degrees of freedom on a dynamically positioned vessel vary with design and number of thrusters as shown in Table 1-8.

Redundant thruster degrees of freedom require that commands in addition to the basic three dynamic positioning com-

TABLE 1-8

Vessel	No. Thrusters (including main propulsion)	Type	Thruster Degrees of Freedom
Eureka	2	Steerable	4
Terebel	2	Steerable	4
Algor	2	Cycloidal	4
Saipem II	4	Cycloidal	8
Challenger	6	Fixed	6
Pelican	7	Fixed	7
Sedco 445	13	Fixed	13

mands be given. For a twin steerable thruster vessel with four degrees of thruster freedom, one degree of freedom is redundant. This unused fourth degree of freedom for the thrusters is the specification of the difference of the thrust of the two thrusters along their line of centers (Figure 1-112). Equal but opposite thrusters along the line of centers produce no change in the surge or sway force or the moment applied to the vessel. This particular redundant degree of freedom has been called bias or anti-oscillation.

The term anti-oscillation stems from use on the Eureka where its use could prevent constant reversals from wave motion. By manual adjustment of the anti-oscillation control, the operating point of the two thrusters could be adjusted until it was far removed from the reversing point of either thruster. Avoiding the reversing point for an azimuthing thruster is very important because of the slow azimuthing (typically 10 to 30 seconds for a 180 degree direction change) and because during the azimuthing change any thrust developed by the thruster is reduced in effectiveness or in error. The fourth degree of thruster freedom for two steerable thrusters is not trivial even when it is set to zero.

For a vessel such as the Sedco 445, which has 13 fixed thrusters and 13 degrees of thruster freedom, the 10 redundant degrees of

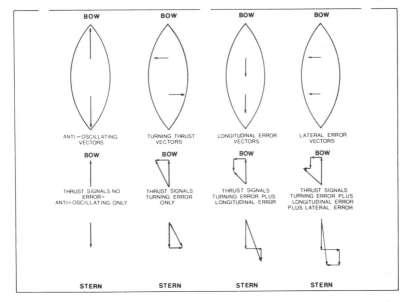

Fig. 1–112 *Vector diagrams for position keeping, 2 Thruster-Four degrees of freedom.*

thruster freedom represent many possible arrangements. Some of the thruster logic arrangements were straightforward and simple such as the statement that "those transverse thrusters operating (out of 11 total) are to equally share in the sway thrust assignment" and when both main screws are operating they shall equally divide the surge thrust assignments.

Developing a pure moment to satisfy the yaw command requires a pair of thrusters to develop a couple. They must also develop equal and opposite thrusts to avoid developing a net surge or sway force. No "moment" or pure moment developing device is known. The really efficient technique for a moment is to develop a couple between the most separated thrusters. For the Sedco 445, consideration was given to putting all of the moment command on the most remote pair of thrusters until they were fully loaded, applying the excess moment to the second most remote pair of thrusters until they saturated, and then selecting the third most remote pair, etc.

This scheme was not adopted because of the lack of appeal of shifting back and forth from thruster set to thruster set with a varying thruster load. Instead, a thruster moment logic was adopted that assigned moment producing thrust couples to all thrusters with a moment arm and weighted linearly proportional to moment arm and rated power.

This moment logic means that all moment producing thrusters share moment commands except that thrusters with twice the moment arm receive twice the moment commands, and thrusters with more horsepower (the main screws) have the command ratio adjusted again after the moment arm apportionment in order to further apportion the moment command according to power. Accordingly, even though the main screws have a short moment arm, their fully powered dynamic position rating of 2,000 horsepower each compared to 800 horsepower for a transverse thruster does result in significant moment generation when only two or three of the transverse thrusters are in service.

On the Sedco 445, the effects of coupling of the surge thrusters (the main screws) and also the coupling of the sway thrusters into the yaw degree of freedom and vice versa is handled by an internal control mode of the dynamic positioning control. This control mode completely reconciles the simultaneous demands of surge, sway, and yaw.

The redundant thruster degrees of freedom basically mean that additional conditions can be set by the system designer. For an array of 11 fixed sway thrusters and two fixed main screws on the Sedco 445, the thruster logic is fairly simple because of the straightforwardness of the fixed thrusters oriented on the control axis of the ship.

With steerable thrusters, however, their flexibility allows some significantly different modes of operation. Consider the simple pattern of the six thrusters shown in Figure 1-113(a). These six thrusters have 6 x 2 or 12 thruster degrees of freedom or 9 redundant degrees of freedom after the positioning requirements of surge, sway, and yaw are satisfied.

If we apply straightforward, equal distribution of surge and sway commands, the simple result of Figure 1-113(b) is obtained. If linear weighting proportional to moment arm thruster

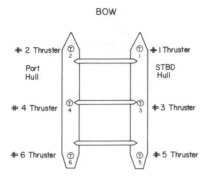

Fig. 1–113A *Lower hull layout, 6 azimuthing thrusters on semisubmersible.*

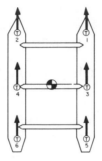

Fig. 1–113B *Equal thrust commands.*

logic is applied then a counter-clockwise moment command will produce the thruster pattern shown in Figure 1-113(c). Notice that using the thrust proportional to length of moment arm logic turns the thrusters so that the thruster vector is at right angles to the moment arm and to the moment center of the vessel. This maximum moment efficiency is derived from the same simple logic used with fixed thrusters of calculating the surge and sway components of thruster using the X and Y coordinates of the steerable thruster.

When the thruster logic of equal division of surge or sway of Figure 1-113(b) is combined with Figure 1-113(c), the combined

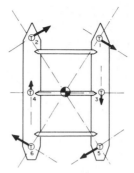

Fig. 1–113C   *Pure turning maneuver-moment arm weighting on thrust.*

maneuver of Figure 1-113(d) is generated. In the thruster pattern of Figure 1-113(d), notice that the moment thrust components and surge or sway thrust components upset any prior thrust equalities and strongly drive some thrusters (as Numbers 2 and 6) to saturation while other thrusters (notably Number 3) are unloaded almost to zero thrust.

Although it is true that the controller can be made to identify saturation and reassign excessive thrust commands, use of internal commands can also be made to manipulate thruster response patterns. These internal commands use the redundant degrees of freedom to write equations governing the group as well as the individual performance of the thrusters.

Consider Figure 1-113(c) which is satisfying the same dynamic positioning commands as that arrangement of Figure 1-113(d). In the thruster logic of Figure 1-113(c), the direction of the resultant surge and sway is detected and all thrusters participate in a joint azimuth command to this direction. Then, with a linear preference to the thruster with the longest moment arm *at this particular azimuth,* moment couples are developed in the direction of the group azimuth.

This mode of Figure 1-113(e) is an economy mode since it strives to have all thrusters delivering thrust in the one direction of the principal surge or sway direction with no wasted thrust. However, of all the thruster logics, that of Figure 1-113(e) is one of the first to have a thruster go into saturation. Figure 1-113(e) thruster logic is "fair weather logic."

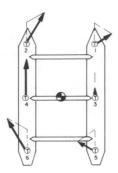

Fig. 1–113D *Thruster pattern of combined motion and turning command, straight forward logic.*

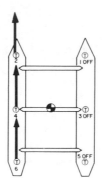

Fig. 1–113E *Thruster pattern of combined motion and turning command, fair weather logic.*

If Figure 1-113(e) is "fair weather logic" then Figure 1-113(f) depicts "power logic." The system of Figure 1-113(e) is designed so that to the equal thruster amplitude response of Figure 1-113(b) moment couples are added at right angles to the resultant surge sway vector. In this manner, with some refinements, the thrusters assume a common, variable magnitude developing moments by deviating the azimuths of individual thrusters.

This mode of operation is called the power mode because it produces the maximum positioning power utilization possible by commanding a common but variable thrust to all thrusters and adjusting an individual thruster's azimuth to satisfy the moment commands.

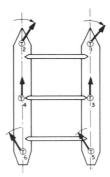

Fig. 1–113F   *Thruster pattern of combined motion and turning command, power logic.*

THE ANALOG SYSTEM

The first dynamic positioning controller, used in the Eureka, is shown in Figure 1-114. Since the loss of a core hole could be accepted from time to time, this first system had no redundancy or backup of any kind. The "joy-stick" in the foreground is used to directly control the surge and sway position when in a manual mode of control. When on automatic control, the "joy-stick" is not in the circuit and the set points in the three turn counting dials determine the position held. Separate three mode controllers provide independent proportional gain, integral action, and derivative settings for surge, sway, and yaw.

Certain aspects of the appearance of the console shown in Figure 1-114 can be attributed to its designer and maker, Hughes Aircraft Company. For this two azimuthing thruster vessel an adjustable bias controller selects bias with wake outward, zero bias, or bias with wake inward. Wake directed outward has been found to add to stability. Wake directed inward, however, is unstable because of the wake action under the hull of the boat and also because of the instability when the two wakes directly impinge on each other. The Eureka utilized a single taut line positioning system.

The Caldrill's analog dynamic positioning controller shown in Figure 1-115 is the first fully redundant controller having two

Fig. 1–114 *Eureka analog dynamic positioning controller. (Courtesy of Hughes Aircraft)*

Fig. 1–115 *Dual analog dynamic positioning controller for Caldrill I. (Courtesy of Baylor Co.)*

separate controllers mounted in opposite sides of a desk sized console. Redundancy was designed into the system with four azimuthing thrusters so that any single bow-stern pair of thrusters could hold position as determined by either controller.

In Figure 1-115, the single "joy-stick" manual control station is seen in the center section in front of the position display. The position display itself is not redundant, as it is not a part of the actual control loop. The Caldrill's permanent position equipment consists of two taut line positioning systems. Installation and a series of successful tests were conducted with a short base line acoustic positioning system complete to the interface with the controller and dynamic positioning by the system.

THE DIGITAL SYSTEM

The digital dynamic positioning controller shown in Figure 1-116 consists of two interconnected units. The control console

SEDCO 445 ASK CONFIGURATION

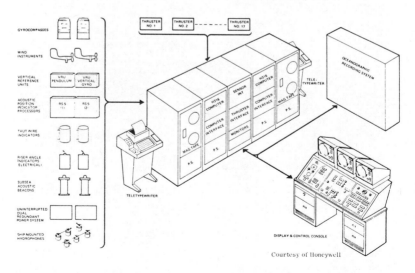

Courtesy of Honeywell

Fig. 1–116 *Dual digital dynamic positioning controller.*

in Figure 1-117 contains two position sensing displays on either side of the console for the two short base line acoustical systems. The left panel contains the gyrocompass heading indication, the surge and sway offset inputs, and the heading set point.

The center section of the control console contains bi-directional ammeters for thruster power units together with individual manual power controls, status reporting lights, and push buttons to transfer control of thrusters from the ship engineer's control center to the control console.

The right panel of the control console is a control and alarm panel for the console operation. Two meters show the amount of total surge and sway force being currently developed as a percentage of that possible with the thrusters currently in operation.

The digital console, Figure 1-118, contains the two computers used in this fully redundant system together with computer/thruster interface, computer/sensor interfaces, power

Fig. 1–117 *Dual dynamic positioning control console.*

Fig. 1–118 *Dual dynamic positioning controller digital console.*

supplies, and a control and alarm section. On one side is a digital magnetic tape data gathering system. Teletype units (not shown) allow troubleshooting and insertion of specific commands to the computer. A very simplified control schematic of such a digital system is shown in Figure 1-119.

COMPARISON OF ANALOG AND DIGITAL SYSTEMS

The analog system has as its advantage a relatively simple, special purpose, dedicated piece of equipment. It can use stan-

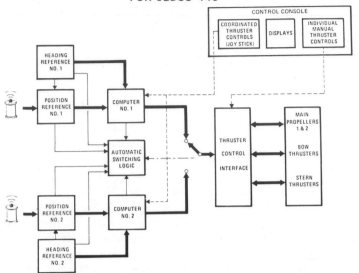

Fig. 1–119 *Dual digital dynamic positioning simplified schematic.*

dard, high quality, interchangeable, industrial three mode controllers. Standardizing the controls allows plugging one common spare into either the surge, sway, or yaw control channel after setting the proportional band, rate, and reset to the channel's known settings.

The advent of the mini-computer has really launched the digital computer into the control field. Unlike the analog system, the digital system has program instructions or software that can be changed to make significant alterations or improvements in performance. In particular, the digital machine may allow use of highly sophisticated wave filters. Also, the digital machine is much more suited to deal with system nonlinearities such as loading thruster A up to saturation and then loading thruster B with that material which thruster A cannot handle.

The digital system is possibly a little more sensitive to control power stability than the analog system. In both systems, care must be taken to avoid power interference with control signals, especially with larger power units being installed today and with the use of alternating current to direct current power conversion systems.

THRUSTERS

Thrusters are labeled as such because it is their purpose to produce thrust in given conditions of current speed and submergence. Thrusters, in general, include all types of devices that can product thrust. This discussion, however, will be limited to devices with propellers—the cycloidal paddle wheel and the water jet.

Of all the types of thrusters, the simplest is the fixed thruster. For dynamic positioning, the fixed type has the obvious disadvantage of not being able to thrust in a specified direction. In a drillship such as the Challenger, the Pelican, or the Sedco 445 there is a line of reasoning that leads to the selection or fixed thrusters. To develop 10,000 lbs. of omni-directional thrust by fixed thrusters it would be necessary to install two 10,000 lb. thrusters at right angles.

In addition to doubling the cost to supply the two fixed thrusters, there is a power demand of 141 percent of that of a single thruster to have both thrusters run to develop the 10,000 lbs. of thrust needed at a direction of 45 degrees. The prohibitive cost and power requirements cited eliminate fixed thrusters for dynamic positioning in all cases except in the case of a ship. The ship, to meet requirements of mobility, will have main propulsion which has generally large horsepower and a large screw.

For dynamic positioning of a ship, very little power is required in the longitudinal direction (because the vessel is a ship). Since the main propulsion already exists through its own need, the reasoning is that for cost and reliability transverse fixed thrusters provide an effective thrusting system when used on a vector resultant with main propulsion.

Due to the reduced power requirements on the longitudinal axis of the ship, the actual maximum needed power on this axis may be only 20 to 40 percent of that on the transverse axis instead of 100 percent as previously assumed. In particular, on the Challenger and the Pelican with their tunnel thrusters, and on the Sedco 445 with its retractable fixed thrusters, the fixed thrusters have a definite application with ships that also have main propulsion.

For a conventional thruster with a propeller, a right angle drive is used and final gear reduction is made in a bevel gear reduction. The vertical high speed driving shaft is the pinion and the low speed gear is attached to the propeller shaft. Power, gear loading, and gear life determine size of this gear. For an 1,800 horsepower unit giving unlimited life, a 49 inch-diameter gear is required. Figure 1-120 shows an 1,800 horsepower 200 rpm thruster.

A precision roller chain makes an excellent means of transmitting power from a horizontal shaft in the thruster pod to the horizontal propeller shaft. A "vee" shaped hollow bracket supports the lower structure and the hollow bracket houses the chain. On an 800 horsepower thruster the central hub is very small. The "vee" structure takes up a little more room than the strut and bevel gear drive housing of a more conventional geared right angle drive. The chain for this thruster is the same as used by drawworks manufacturers.

Fig. 1–120   *1,800 hp azimuthing thruster. (Courtesy of Baylor Co.)*

A fixed pitch propeller is, of course, the simplest type. For dynamic positioning, this fixed pitch propeller requires a variable speed to obtain variable thrust and reversing of the direction of the propeller to change direction of thrust. Both forward and reverse characteristics of the propeller must be considered. The 800 horsepower fixed pitch propellers of Figure 1-121, for example, showed only six seconds from full on forward, through stop, to full on reverse.

A controllable pitch (CP) propeller is appealing for the dynamic positioning control viewpoint because the propeller runs at constant full speed with control power going to control the pitch of the unit from full ahead, through neutral, to full ahead (Figure 1-122). An example of the use of controllable pitch (CP) is the Pelican, which uses five CP 1,500 horsepower transverse thrusters and two 3,000 horsepower CP main screws.

Open type propellers, usually identified by their round propeller tips develop thrust somewhere in the range of 20 to 30 lbs.

Fig. 1–121 *Twin 800 hp bevel gear fixed pitch Kort thrusters. (Courtesy of Baylor Co.)*

Fig. 1–122   *800 hp controllable pitch tunnel thruster. (Courtesy of Baylor Co.)*

of thrust per horsepower. Generally, reducing the propeller area horsepower loading gives greater specific thrust in lbs. thrust per horsepower.

The tunnel type thruster, as used on the Challenger and the Pelican, has the very great advantage of being within the hull of the vessel. Reduced performance in the range of 20 lbs. of thrust per horsepower is realized because of the losses in the length of the tunnel and the pressure losses on the discharge hull side.

The Kort nozzle is a special nozzle and propeller developed by Kort of Germany. The basic design allows for lower speeds to develop extra thrust. The propeller and nozzle shown in Figure 1-121 are typical Kort nozzles. A Kort nozzle, properly designed, will have a thrust of between 30 and 40 lbs. per horsepower.

Water jet thrusters are not really considered in dynamic positioning because of their very low thrust efficiency, estimated at below 10 lbs. per horsepower.

Azimuthing thrusters have means with which to direct full thrust at any commanded azimuth. Since azimuthing thrusters can change direction of thrust by azimuthing, they need not be reversing thrusters. By changing direction by azimuthing instead of reversing the thruster prop, a loss of 8 to 12 lbs. per horsepower for the reversing thrusters is avoided. Furthermore, the azimuthing thruster is more efficient in realizing the thrust vectors not aligned with one of the principal axes of the thrusters.

Typical azimuthing time for a 180 degree reversal on a thruster is 30 seconds. On a 1,800 horsepower open type propeller, as shown in Figure 1-122, a 20 horsepower azimuth drive motor is used. On a 2,000 horsepower thruster with a Kort nozzle, two 40 horsepower azimuth drive motors are used because of the large rudder effect of the Kort nozzle. Figure 1-123 shows a 300 horsepower azimuthing, retractable thruster.

The Voith-Schneider type of thruster shown in Figures 1-124 and 1-125 is a distinct category of thruster. Manufactured by

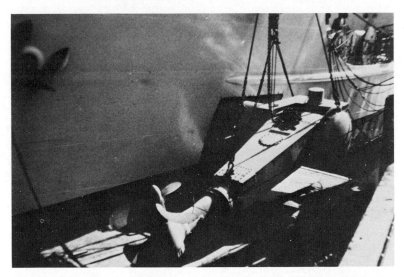

Fig. 1–123   *300 hp azimuthing retractable thruster. (Courtesy of Baylor Co.)*

Fig. 1–124   *1,100 hp Voith Schneider thruster. (Courtesy of Baylor Co.)*

Fig. 1–125   *2,550 hp Voith Schneider thruster. (Courtesy of Baylor Co.)*

Voith for the Saipem II, the thruster consists of a flat, large-diameter rotor or platform that is horizontally flush with the bottom of the hull. Extending vertically downward from a pitch circle near the outside of this disk are five propeller blades. For the 2,550 horsepower thruster of Figure 1-125, the rotor turns the five blades at a speed of 48 rpm. A mechanical system above the rotor introduces cyclic pitch variations in the five propeller blades to develop a given magnitude of thrust in a given direction.

The cycloidal propeller does not develop thrust by "paddling" as one might expect. Rather in develops thrust by skulling, by setting the pitch so that a blade develops hydrodynamic lift as it crosses the direction of travel twice each revolution of the rotor, once on the side in the direction of travel and once on the side away from the direction of travel.

Control of the thrust magnitude and direction is made in the longitudinal (surge) and transverse (sway) axis in the cycloidal unit by two X (sway) and Y (surge) hydraulic servo motors. The net vector resultant is performed by the thruster. Speed of response is very fast (about 4 seconds) for changing the pitch from full ahead to full astern, from hard starboard to hard port, or to any commanded direction.

The cycloidal propeller also runs at a constant speed (it is a special configuration of a controllable pitch propeller). The magnitude and direction of thrust can be varied without steps and with high precision while the prime mover is moving at a constant speed. Furthermore, maximum thrust is available in any direction.

The performance and efficiency of the cycloidal propeller is strongly affected by hull design. For optimum performance, other design requirements may have to be sacrificed to meet the cycloidal propeller requirements. The second concern about the cycloidal propeller is that it is a large precision piece of machinery with many moving parts. The life of seals and bearings is a matter of concern, for instance.

Thruster power sources vary from the direct driven azimuthing thrusters driven at a variable speed by directly coupled 300 horsepower diesel engines to a 3,000 horsepower synchronous motor driving a controllable pitch propeller at a constant speed. In most cases of dynamic positioning today, an electrical drive unit will be used for convenience and control and the ship will probably be diesel electric powered.

A particular class of power system for propulsion and thrusters has been popular. In this popular class the propulsion motors and thruster motors are selected to be the same as the direct current motors for the drilling equipment (Figure 1-126). The typical drilling motor is a series or shunt excited DC locomotive traction motor rated at 750 horsepower continuous or 1,000 horsepower intermittently. The motors have a variable speed of 0 to 1,100 or 1,200 rpm and can be rated by the American Bureau of Shipping for 750 or 800 horsepower for main propulsion service.

These motors are ideally suited for variable voltage, variable speed control driving, fixed pitch azimuthing, or reversing thrusters. Voltage control can be by the older generator field control of a DC motor generator set or by controlled rectification of an alternating current bus voltage to give the variable DC voltage.

Fig. 1–126 *DC drilling motor (GE 752 motor).*

A larger DC motor of 2,250 horsepower, shown in Figure 1-127, has been used in powering the thruster of Figure 1-120. Similar controlled rectification of AC power to give a variable DC voltage is used to power this thruster.

In another thruster power design, a 2,500 horsepower alternating current induction motor is driven by a single, dedicated 2,100 kw. 4,160 volt generator. Speed control to 50 percent speed is performed by slowing the generator down to 30 hertz while proportionally reducing the voltage. Final speed reduction is obtained by holding the frequency constant and reducing the voltage gradually to a minimum of 10 percent of full voltage. By this combination of variable frequency and variable voltage, power to the thruster is varied over a 100 to 1 ratio.

Many other power choices are possible and it is notable that a fixed pitch thruster will require a variable speed power source. If it is a fixed pitch fixed thruster, reversing will also be required. Controllable pitch, including cycloidal thrusters, can be run at a constant speed so that synchronous or high efficiency, low slip AC induction motors may be applied. Coordination, or the use of the same power units in the thrusters or propulstion as used in

Fig. 1–127   *2,000 hp DC motor (GE 608 motor).*

the drilling equipment (which was done on the Challenger and Sedco 445), may be a worthwhile objective.

The thruster will be housed either in the hull or it may be in a pod attached or inserted into the hull. Access to the inner compartment of the pod is necessary for maintenance and repair of the thruster. Following is a partial list of services that may be required or desired for this space:

1) Bilge pump,
2) Bilge alarm on bridge,
3) Sound powered telephone into vessel's system,
4) General alarm bell,
5) Motor space heaters,
6) Elevation access for semisubmersibles.

The general alarm bell is needed to avoid being trapped below if there is a general alarm. Lights, motor space heaters, and bilge pumps should be on the ship's service power supply to operate even when the propulsion or thruster is turned off. In addition, the pod should operate at less than 50 degrees C, preferably 40 degrees C, so that maintenance can be performed while the unit is operating.

MODEL TEST

Model tests are a must for the design of a dynamic positioned vessel for large sums of money are being committed in an area where there is little existing data or experience. Major model basins have the technology, people, technique, and equipment to perform directed work necessary to supply dynamic response of the vessel to wave, current, and wind and also thruster performance.

Model scale is chosen for the relative need for accuracy. In ships, for instance, large-scale wax models of 14 to 20 ft. in length are used with powered scale propulsion to detail the expected developed speed versus power for calm water and various sea conditions. For maneuvering tests, a smaller scaled powered model may be radio controlled in a sea keeping basin where the vessel's maneuverability is checked in the face of various obstacles with waves of differing heights and directions combined with winds.

Tests of this same scale model may be performed in oscillation tests in which the complete spectrum of the model's response to waves of all periods and all angles of attack is determined. The model's resistance to currents at different angles of attack is measured and wind tunnel force studies generate the model's resistance to wind. A typical plot of forces and horsepower requirements on a ship are shown in Figure 1-128.

Thruster details and studies of developed thrust can be figured with large-scale powered models of a scale of about 1:12. These tests determine thrust versus horsepower at varying current speeds and directions. If interference of an adjacent thruster or possible pressure drag on a hull is suspected, these items are checked with this model. On the Sedco 445, for example, these model tests resulted in moving thrusters out of alignment with each other across the ship because the high water flow at the inlet of the down wash thruster from the prop wash of the other thruster very seriously degraded the performance of the down wash thruster. The solution was to put all of the thrusters on one side of the ship at the bow and stern. A high low pressure area drag or thrust reduction was also shown by the prop wash curving around and following the side of the ship.

The vessel environmental reaction tests (to determine the forces on the vessel due to wind, wave, and current) together with the thruster performance tests (or developed thrust versus input horsepower, thrust command, and current speed and direction) are usually combined in a simulation computer. In this computer, where vessel performance (as defined by model test results) meets with the thruster performance, a computer model of the control system or the actual control system can be used to close the control loop in a simulation study.

In this simulation study, all of these factors are weighed together and output data and curves are generated showing simulated vessel performance. These simulated vessel performance curves are compared with contract requirements, and in many cases a return to some part of the model tests may be necessary to improve performance and generate new model test input data to perform a second simulation run.

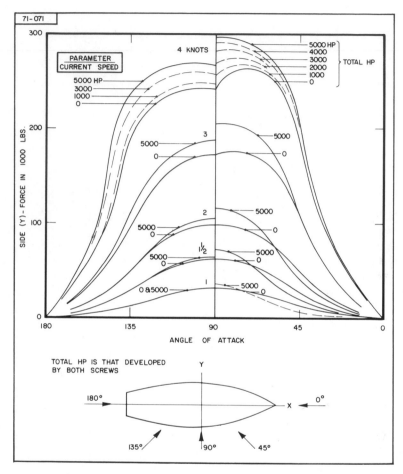

Fig. 1–128   *Forces and power requirements from model tests. (Courtesy of Sedco)*

Model test data of Figure 1-129 show that performance always falls off with current speed except when the current is pushing against the propeller. Also, note the difference in inward running versus outward running. To obtain thrust that is nearly equal in both directions while trying to maximize thrust in both directions entails difficult design compromises in the Kort nozzle.

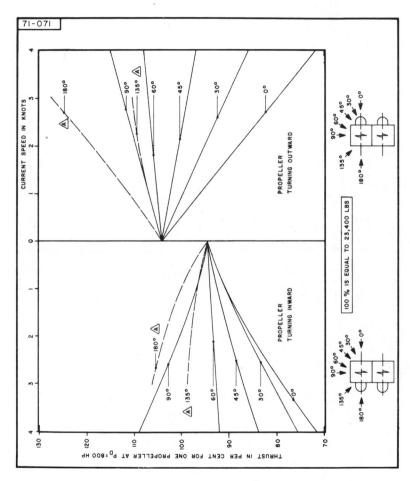

Fig. 1–129   *Performance of reversible bevel gear Korted 800 hp thruster.*
(Courtesy of Sedco)

Part II

Motion Compensation
and Marine Risers

6

Tensioning Systems

William L. Clark
Hydril Company

Marine Riser Tension Requirements

A marine riser operating on a floating drill rig will fail (collapse) in water depths greater than 200 to 300 feet if it is not partially or completely supported. The marine riser is attached to the sea bottom via the blowout preventer BOP stack. Thus, it cannot be firmly attached to a floating drill rig. The support must come from axial tension applied to the top of the riser and/or buoyancy along the length of the riser. Tension controls the stress level in the riser pipe and affects the riser straightness during drilling operations. As rise size, water depth, sea conditions, mud weights, etc. increase, the axial tension requirements for providing proper support also increase.

Calculating the amount of axial tension required is a complex beam deflection problem with a number of variables. This mathematical solution should be applied to all field applications. However, a "rule of thumb" to determine a rough order of magnitude (ROM) utilizes the weight of riser pipe in water, weight of drilling fluid in water, and a safety factor. The ROM nominal tension is:

T nominal = (weight of riser in water+ weight of drilling fluid
in water) x 1.20.

This level of tension will maintain the bottom joint of the riser in positive tension as it exceeds the riser weight.

Example:
a. The weight of 20 inch OD riser with ½ inch wall in water is
    equal to 146 lbs. per foot
b. Drilling fluid weight = 18 lbs. per gallon; weight in water
    equals 147 lbs. per foot
c. Nominal tension in 800 foot water depth = (800 ft.) (147
    lbs. per ft. + 146 lbs. per ft.) x (1.20) = 281,280 lbs.

Standard tensioner sizes utilized today are 60,000 lbs. or
80,000 lbs. per tensioner. These units are utilized in systems
containing 4, 6, or 8 tensioners. Thus, the application above
would require a minimum tensioner system comprised of six
60,000 lb. or four 80,000 lb. units.

Riser tensioners are operated in pairs so that the two tensioners connected on diagonally opposed sides of the riser are always at the same tension level. Additional pairs of tensioners are
used to achieve both redundancy of tension availability and
high levels of tension by utilizing all the tensioners on the riser
at the same time. New rigs, presently being designed for large-
diameter risers and/or deep water depths, commonly use six or
eight 80,000 lb. riser tensioners and a few are designed to operate with ten units. A well designed system will provide the
maximum required tension level with at least one tensioner out
of service.

Basic Tensioner Specifications

This list delineates the basic requirements of any good tensioner design. The units must have at least these basic
capabilities in order to function in their application.
a. Tension Capacity
    The tension capacity is determined by the ultimate or
    maximum riser tension requirement. Several tensioner units
    are utilized in all riser tensioner systems. Multiple tension-
    ing units provide larger capacity tensioner systems and en-
    sure some operational redundancy or safety capacity to
    permit operation with a unit which is down for mainte-
    nance.
b. Wireline Travel
    The tensioner wireline travel or motion compensation
    capability must exceed the maximum vessel heave expected

to occur while the riser is connected to the wellhead. The tensioner motion capacity must not only exceed vessel heave but must also account for tidal motion, connection adjustment, and changes in vessel ballast position. Excessive travel requirements will normally be detrimental to overall operational efficiency of any tensioner. It is therefore desirable that a design optimize rather than maximize the travel capability.

c.   Speed Response Capability

The tensioner must have the capability to respond at the maximum peak response of rig vertical heave motion. This response must equal or exceed the instantaneous maximum vertical speed of the vessel heave, which exceeds the average vertical speed of the vessel.

The maximum velocity (V) can easily be calculated (assuming a sine wave):

$$\frac{\text{Heave (ft.) x pi (k)}}{\text{period (sec.)}} = \text{Maximum Velocity (ft./sec.)}$$

Example: heave= 16 ft.; period = 16 sec.; V max. = 3.41
         ft./sec.; V avg. = 2 ft./sec.

Notice the difference between the maximum velocity and the average velocity.

Basic Tension Sources

The counterweight was the first technique utilized to apply tension to the top of marine risers. The weight was hung from a wire rope which was reaved up over wire rope sheaves and down to the top of the riser pipe. The tension was equal to the counterweight (Figure 2-1).

This technique proved to be inefficient and unsafe as the tensioner requirements (tension and travel) increased. Later designs improved this technique, but were unable to overcome the disadvantages. There are a few of these systems still in operation, but for several years no new rigs have been manufactured using counterweight tensioning.

Presently, there are two techniques being applied to obtain riser tension by buoyancy. One technique applies flotation material, which is molded to shape and is strapped to the riser pipe

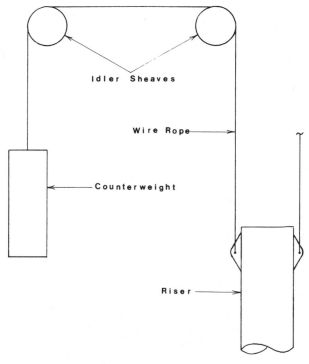

Fig. 2–1   *Counterweight.*

(Figure 2-2). A second technique applies concentric air-filled
chambers to the riser pipe and obtains its buoyancy by en-
trapped air (Figure 2-2).

Both techniques reduce the weight of the riser system in
water. This weight reduction reduces the required axial tension.
Most (if not all) of the present riser applications apply buoyancy
to no more than 95% of the weight of the riser in water. They also
utilize other top tension devices (tensioners) to provide the
positive tension requirement. Utilization of buoyancy on a
marine riser does not eliminate the need for tensioners, but it can
significantly reduce the amount of tension required and thus
reduce the quantity or size of tensioners that must be applied.
There are subsea system designs being evaluated which utilize
positive buoyancy, but they too apply top tension.

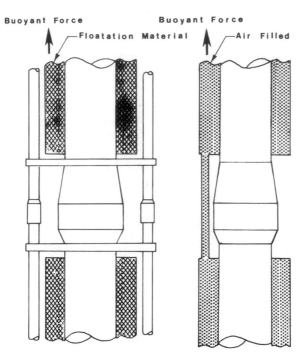

Fig. 2–2 *Buoyant riser.*

A pneumatic spring is the predominant technique utilized to obtain top tension in the marine riser. This system uses compressed air for energy. The spring force of the compressed air is translated into linear motion by hydraulic-pneumatic cylinders or rams. The spring force is transmitted from the cylinder/ram by wire rope to the top of the riser pipe.

Explanations of the various ways this technique is applied are presented in the next section.

Pneumatic Spring Tensioners

All presently available tensioners create linear force in a hydraulic-pneumatic cylinder/ram by application of high pressure compressed air. This linear force is transmitted by a wire rope to the top of the marine riser to provide axial tension.

There are several techniques currently being used. The main techniques and requirements are:

a.   Ram Cylinder—Compression Loaded

This cylinder is a single acting ram as shown in Figure 2-3. Pressure acting on an area, which is determined by the diamater of the rod, forces the rod to extend. The extension force is equal to the pressure multiplied by the square area of the rod. For example a 16 inch bore cylinder with a 14 inch diamater rod would operate with pressure on the 154 square inches of the rod. At 2,000 psi the extension force would be 308,000 lbs.

To provide lubrication of bearings and pressure packing and to provide a means for instant pressure shut-off for safety control, an air/oil interface is applied. The high pressure air is connected to an air/oil accumulator and the oil side of the accumulator is connected to the cylinder. The ram is pressured by oil.

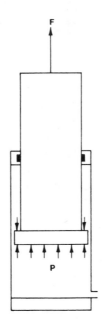

Fig. 2–3   *Hydraulic ram.*

The extension force of the ram is directly proportional to the pressure. Therefore, the operating force (tension) level is set by controlling the pressure in the cylinder. The pressure levels range from a few hundred pounds per square inch to approximately 2,500 lbs./square inch.

b. Double Acting Cylinder—Compression Loaded

This hydra-pneumatic cylinder is a double acting cylinder as shown in Figure 2-4. Pressure acting on the full area side of the piston forces the rod to extend. The extension force is equal to the pressure multiplied by the square area of the piston. For example a 14 inch bore cylinder would operate with pressure on the 154 square inches of the piston. At 2,000 psi the extension force would be 308,000 lbs. To provide lubrication of the bearings and high pressure packing an air/oil interface is connected to the high pressure side of the cylinder. High pressure air is connected to the air side of the accumulator and the oil side of the accumulator is con-

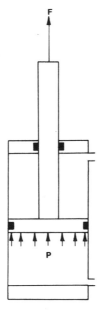

Fig. 2–4   *Hydraulic cylinder, compression type.*

nected to the cylinder. Therefore, the cylinder is pressured with oil rather than air.

The extension force of the cylinder is directly proportional to the pressure. Therefore, the operating force (tension) level is set by controlling the pressure in the cylinder. The pressure levels are from a few hundred psi to approximately 2,500 psi.

A low pressure air/oil reservoir is connected to the rod end of the cylinder to provide lubrication of the rod packing and bearings, the rod end of the cylinder piston packing and bearings, and for safety control.

c. Single Acting Cylinder—Tension Loaded

This cylinder is a single acting cylinder as shown in Figure 2-5. Pressure acting in the rod side of the piston forces the rod to retract. The retraction force is equal to the net piston area multiplied by the pressure. For example, a 12.5 inch bore cylinder with 6 inch diameter rod would have

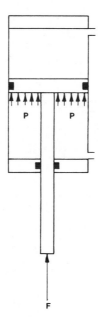

Fig. 2–5 *Hydraulic cylinder, tension type.*

a net rod end area of 100 square inches. At 3,260 psi the retraction force would be 308,000 lbs.

To provide lubrication of bearings and pressure packing, and to provide a means for pressure shut-off for safety control, an air/oil interface is applied to the high pressure side of the cylinder. High pressure air is connected to the air side of the accumulator and the oil side of the accumulator is connected to the cylinder. Therefore, the rod side of the cylinder is pressured with oil.

The retraction force of the cylinder is directly proportional to the pressure. Therefore, the operating force (tension) level is set by controlling the pressure in the cylinder. The pressure levels are from a few hundred psi to approximately 3,500 psi.

Wire Rope Reeving

The cylinder or ram force is transmitted to the top of the marine riser through wire rope. In order to translate the linear cylinder motion into wireline tension the cylinder is fitted with wire rope sheaves at each end.

Design evaluations have shown that the optimum number of line parts is four. This connection means that for each unit of stroke by the cylinder it will provide four units of stroke in the wire rope. This reeving also means that each unit of force in the cylinder is reduced by a factor of four. In other words, the cylinder travel is multiplied by four and the cylinder force is divided by four, as seen in the wire rope. This mechanical connection affects cylinder sizing. It was selected to optimize cylinder size and the pressure at which the cylinder must be operated to achieve the required tension level. For example, a cylinder force of 308,000 lbs. would provide 77,000 lbs. of wireline tension, and 12.5 feet of cylinder stroke will deliver 50 feet of wireline travel.

Two wire rope sheaves are connected at the rod end of the cylinder and two sheaves are connected to the fixed end of the cylinder. In the tension loaded cylinder the fixed sheaves are not mounted on the cylinder but are away from the cylinder as shown in Figure 2-6. Wire rope is reeved around the cylinder

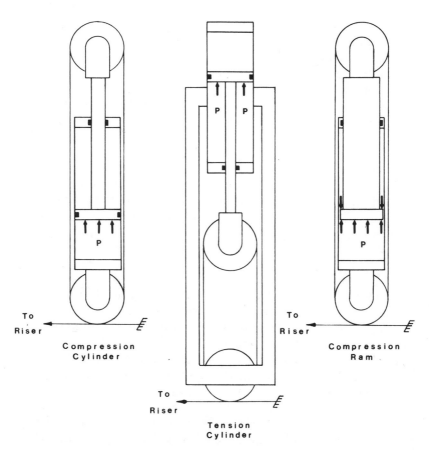

Fig. 2–6 *Tensioner variations.*

sheaves to form a loop around the cylinder. Expansion force of
the compression loaded cylinder or retraction force of the tension loaded cylinder attempts to expand the wireline loop.
Thus, the force develops a pull or tension in the wire rope.

One end of the wireline is connected to the tensioner or rig
and the other end is reeved across an idler (diverter) sheave and
to the top of the marine riser (Figure 2-7). Force in the cylinder is
thus translated to wireline tension and then to tension in the
riser. Air pressure is maintained at an approximately fixed level
to maintain the desired tension level in the riser. As the rig
heaves upward, the top of the marine riser remains fixed (as it is

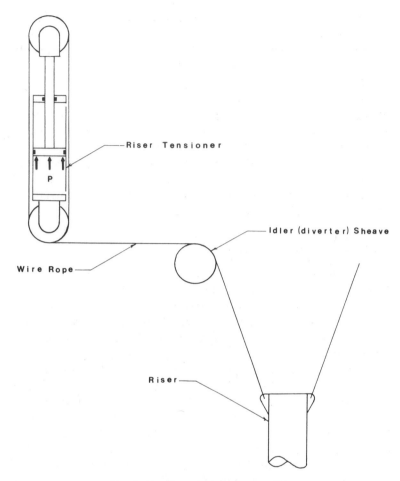

Fig. 2–7   *Riser tensioner reeving.*

fixed to the earth by being connected to the wellhead). Thus, the wire rope, which is fixed to the rig and to the riser pipe, tends to pull more taut. This pull on the wire rope compresses the compression loaded cylinder (retracting the rod) and thus lengthens the effective length of wire rope. By compressing the wire rope loop it maintains the tension level as determined by the air pressure contained in the cylinder.

When the rig heaves downward the wireline tends to slacken, and the compression cylinder expands to take up the slack and maintains the present tension level. The tension loaded cylinder operates the same as above except that the cylinder moves in the opposite direction to maintain a taut wire rope. This is due to the retraction force on the cylinder rod rather than expansion force.

There is a definite relationship between the sheave design, sheave diameter, and the wire rope design. The wire rope specified by the tensioner manufacturer should be utilized to realize the best wire rope life. It is important not only to use the correct wire rope diamater but also the correct construction, e.g. 6 x 37 IWRC IPS, 8 x 41 IWRC IPS and so forth. The wire rope must be flexible to resist metal fatigue due to the average 6,000 reversal bending cycles per day over the sheaves incurred during normal applications.

Pressure Control in Cylinder

Pressure in the cylinder determines the wireline tension and thus the riser tension. To maintain the tension at or about a preselected level the air pressure must be maintained at a fixed pressure level. The technique which is used in all presently available tensioners is compressed air stored in high pressure reservoirs or air pressure vessels (Figure 2-8).

Air is compressed into the air pressure vessel by high pressure air compressors to a level which will pressure the tensioner cylinder to the desired tension. The air pressure vessel and the tensioner cylinder are directly interconnected through adequate size piping (incorrect piping sizes will adversely affect the system operation) so that the cylinder and the air pressure vessel are at the same pressure. The extension force of the cylinder is translated into wireline tension, which is connected to the top of the riser pipe to provide riser tension.

As the drill rig heaves upward and compresses the cylinder by pulling on the wire rope and causing the cylinder to retract (paying out more wire rope to maintain the preselected tension level), the air in the cylinder is compressed into the air pressure vessel. When the drill rig heaves downward, the cylinder expands (to take up on the wire rope). Air from the air pressure vessel then expands into the cylinder, maintaining the cylinder pressure and wire rope tension level.

The amount of air pressure change and tension variation due

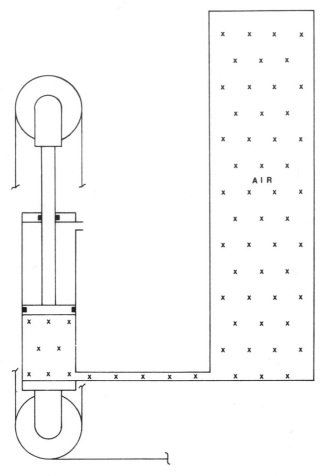

Fig. 2–8 *Compressed-air system for wireline tensioner.*

to the expansion and compression of the air during stroking of the tensioner cylinder is determined by the ratio of the cylinder volume and air pressure volume. A small air pressure vessel would result in large tension variations due to rig heave. Very large air pressure vessels would virtually eliminate tension variations due to rig heave (compression of air) but would be too heavy, large, and expensive. An optimum ratio between cylinder size and air pressure vessel size has been selected by each tensioner manufacturer.

The compression of air into the air vessel is a polytropic cycle and can be calculated by the formula:

$$P_1V_{1n} = P_2V_{2n}, \text{ where } n = 1.1$$

It has been found by experience and testing that this value for n most closely approximates the condition that occurs during normal heave operations.

Control Panel

Control panels are primarily valve manifolds. Most of the valves utilized are simply off-on valves and are operated in the full open or closed position. During a normal operation no new air is compressed into the system nor is it released to the atmosphere except to change the tension level and to replace air which has leaked off. This air replacement normally occurs only on a 6 to 24 hour basis. These functions and adjustments are normally accomplished manually.

Safety Control

It is imperative that any pneumatic tensioning device be fitted with and have as an integral part of its design a safety shut-off device. Due to the possibility of wireline breakage during normal tensioner operation it is always possible that severing or inadvertent breaking of the wireline could occur. When a wireline breaks, the cylinder, pressured by compressed air, is now free to accelerate as fast as the air can expand. This expansion can occur at rates of 15 to 20 gravitational forces. If the cylinder is allowed to freely accelerate to its maximum potential velocity it will virtually self destruct. In analyzing any tensioner and/or pneumatic compensation device it is highly recommended that a close look be taken at the safety systems employed in the design.

Guideline Tensioners

Wellhead guidelines must be taut to be effective. In order to maintain guidelines taut at a preselected tension level, hydra-pneumatic tensioners are applied to each of the four wellhead guidelines and normally to the blowout preventer control lines

(Figure 2-9). Guideline tensioners operate exactly the same as riser tensioners and are designed the same, except they are smaller in size.

Normal operation of guideline tensioners sets them at high tension levels for guiding equipment down to the wellhead, i.e., 10,000 to 16,000 lbs. being common. After landing operations are complete, the tension levels are reduced to a standby level such as 2,000 to 3,000 lbs. in order to reduce the wear and fatigue on the tension and the wire rope.

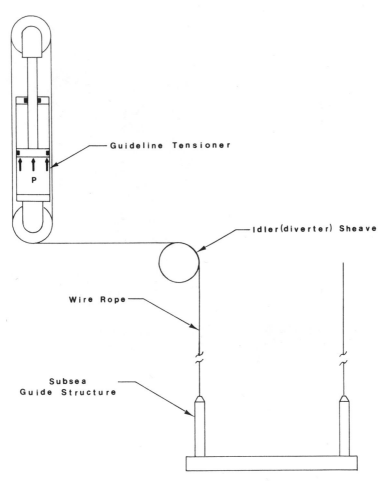

Fig. 2–9 *Guideline tensioner reeving.*

Riser tensioner (60,000 lb. unit) operating in the North Sea at 30,000 lbs. (Courtesy of Rucker Control Systems)

Riser tensioner (80,000 lb. unit). (Courtesy of Western Gear Corp.)

Riser tensioner (80,000 lb. unit). (Courtesy of Vetco.)

7

Considerations
for Marine Risers

Dr. D. Bynum, Jr. and
Ramesh K. Maini
ETA Engineers, Inc.

A riser can be simply described as a conduit from the platform to the ocean floor which transmits the drilling mud and serves as a guide for the drill string. There are two broad classes of risers, those used for exploratory drilling operations and those used for production operations.

In exploratory drilling, the riser tube running from the vessel down to the wellhead is normally called a riser while the first casing joint running from the wellhead slightly above the mudline down through the template to 30-100 feet below the mudline is called a conductor or surface pipe. In production operations, the riser tube running from the platform deck down to the wellhead is also sometimes called a conductor.

In exploratory drilling operations conducted with a jack-up rig, the riser runs from the drilling deck down to the wellhead (Figure 2-10). The drilling fluid is pumped down the drill string. The mud returns in the annulus between the string and the riser in the water or between the string and the casing for that part of the drill string below the subsea wellhead. Ordinarily, when oil is struck in exploratory drilling the well is tested and capped and the rig moves on to another drilling site. In most areas the regulations require that all equipment be removed from the ocean floor. The riser string can then be reused.

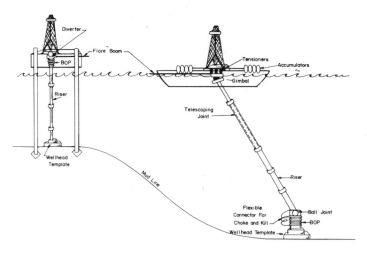

Fig. 2–10   *Marine riser systems for exploratory jack-up and floater drilling operations.*

For floating drilling operations using a drillship or semisubmersible, more hardware is required to accommodate the
vessel's heave, sway, and surge (Figure 2-10). Attention in this
article is directed toward floating drilling operations since the
riser-related problems are more complex.

BASIC COMPONENTS OF AN
EXPLORATORY RISER SYSTEM

Figure 2-10 illustrates the arrangement of the riser system. It is
important to note that the riser is not a simple piece of equipment but rather a complex system of component parts. A list of
the sources for components of basic riser equipment and the
manufacturers of each component has been prepared (Table
2-1). Many manufacturers produce individual parts of the subsea riser system but only a very few supply the entire equipment
package.*

*Table 2-1 was compiled from the 1974 *Composite Catalog* and represents only
those manufacturers included therein. The authors would of course be grateful to
hear from any supplier of riser equipment not listed.

TABLE 2-1

Standard Riser Equipment

Manufacturer	Headquarters	Joint Coupling	Ball Joint	Telescoping Joint	Tensioners	Diverters	Motion Compensators	Subsea BOP Equipment	Subsea Wellhead Equipment	Flexible Riser–BOP Connector
Aeroquip Corp (Barco)	Jackson, Mich.		x							
A-Z Int'l Tool Co.	Houston, Tex.			x						
Cameron Iron Works	Houston, Tex.	x	x	x			x			
FMC Corp (Chicksan)	Houston, Tex.								x	x
Gray Tool Co.	Houston, Tex.	x							x	
Hydril	Los Angeles, Cal.									
Hydro Tech	Houston, Tex.	x						x		
Johnston Div.	Houston, Tex.						x			
National Supp. Co.	Houston, Tex.	x	x	x					x	
Ocean Sci. & Eng.	Long Beach, Cal.		x							
Regan Forge & Eng.	San Pedro, Cal.	x	x	x		x		x	x	
Rucker Control Sys.	Houston, Tex.	x		x	x	x	x	x	x	
Stewart & Stevenson	Houston, Tex.							x		
Vetco	Ventura, Cal.	x	x	x	x	x	x	x	x	
Western Gear Corp.	Houston, Tex.				x		x			

The riser string for a floating exploratory vessel is usually made up of 50 ft. long joints which can be stacked and stored on deck during transit from one drilling location to another (Figure 2-11). The ends of each joint have quick-disconnect couplings permanently attached to the joint. The telescoping joint which is at the upper end of the riser string is usually designed for a maximum heave of between 15 and 30 feet.

A constant force tensioner system is attached to the top of the fixed outer barrel of the telescoping joint to provide enough axial force in the riser string to prevent buckling. The outer barrel and the riser string have lateral movement with vessel surge and sway but essentially no vertical movement with vessel heave. The vessel and the inner barrel of the telescoping joint move together vertically with vessel heave. The optimum ten-

Stacked Riser Joints   Riser Tensioner

Fig. 2–11   *Pacesetter III in transit showing stacked and stored riser joints on deck and riser tensioners mounted on derrick structure. (Courtesy Rucker Control Systems and Western Oceanic, Inc.)*

sion is a function of water depth and operating conditions (mud weight, etc.), as shown in Figure 2-12.

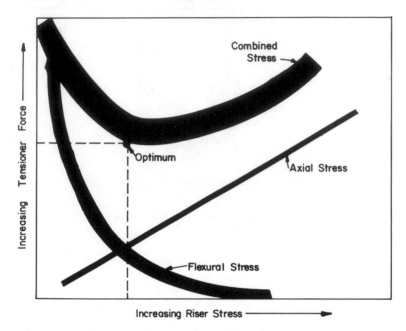

Fig. 2–12 *Optimization of riser tensioners.*

Ball joints (Figure 2-13) on each end of the riser allow for rotation in any direction up to about 7-10 degrees. Actually, only a few operators insist on two ball joints, which offer more reliability than a single ball joint, because the use of two ball joints incurs greater costs and greater running time. The usual arrangement for floating drilling operations (Figure 2-10) is a gimbal under the drilling deck and one ball joint attached to the top of the subsea BOP stack, which sits on the wellhead. The wellhead attaches to the base template that is set with the conductor type at the beginning of the operation.

As drilling progresses, casing supported by casing hangers in the wellhead is placed in the well bore. At intermediate depths, another casing string of smaller diameter is set inside the first casing string from other casing hangers in the wellhead. The

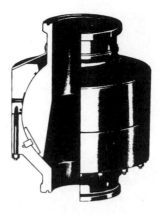

Fig. 2–13 *Ball joint. (Courtesy Regan Forge & Engineering Co.)*

depth and number of the various sizes of casing strings depend upon geological conditions.

Riser Coupling Joints

The first riser systems had the choke and kill lines strapped to the riser pipe. Pipe handling problems with this system proved to be extremely time-consuming. Most riser systems now use integral choke and kill lines (Figure 2-14) which are permanently attached to opposite sides of the riser and have their own connectors. When the riser joints are stabbed and quick-connected the receptacles allow the choke and kill lines to be stabbed and automatically connected at the same time.

The BOP requirements have been the decisive factor in determining the diameter of the riser pipe and hence the riser wall thickness, required tensioner force, etc. In early drilling operations, a two stack 20 inch BOP was used until the first few strings of casing were set. The 20 inch BOP was then changed out for a 13⅜ inch BOP for making the rest of the hole. Although this procedure is still used, and preferred by some, the single stack is growing in usage. The first single stacks were 16¾ inch and required a 17½ inch underreamed hole for setting 13⅜ inch casing. Because of dissatisfaction with underreaming, a single 21¼ inch BOP was then used.

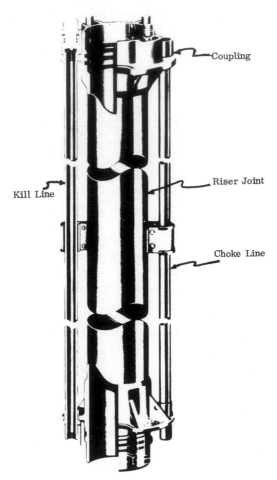

Fig. 2–14   *Riser joint/integral choke and kill lines. (Courtesy Regan Forge & Engineering Co.)*

The heavy weight of these systems, which sometimes exceeded 200 tons, overstressed the system in rough weather. The proposed "North Sea riser system" is a single 18¾ inch stack which weighs half as much as the 21¼ inch system, yet does not require underreaming for the 17½ inch casing hole.

Choke and Kill Lines

The choke and kill lines run from the deck along the riser string down to the wellhead. At the lower riser ball joint there are various schemes, such as looped pipes, to get the required flexibility in a jump line arrangement running from the bottom of the riser string (top of ball joint) around the ball joint to the BOP stack. The choke and kill lines control kicks in order to prevent them from developing into blowouts.

When a potential blowout is detected, mud is pumped down the kill line at the BOP stack to restore pressure balance in the hole. When excess gases occur the bag and ram type BOPs are closed around the drill string. The gas is relieved at the choke manifold on the BOP stack by running up the choke line on the riser string. As the gas expands it proceeds up the choke, displacing more mud and traveling faster as the gas bubble or gas-entrapped mud approaches the mean water line. Without the choke, the gas would push out the annulus mud between the drill string and the riser control from the weighted mud would be lost.

Telescoping Joint

There are two basic types of telescoping joints used with marine risers. The constant tension system (remote axial tensioning system) is most often used because maintenance is easier (Figure 2-15). This method uses a linkage system at the base of the drilling floor to maintain equal force on the several wire ropes attached to the outer barrel of the telescoping joint.

An alternate design of telescoping joint uses the direct axial tensioning method. This is a procedure where the seals and guide rings on the telescoping joint are designed to compensate for internal pressure so that the telescoping joint has the dual

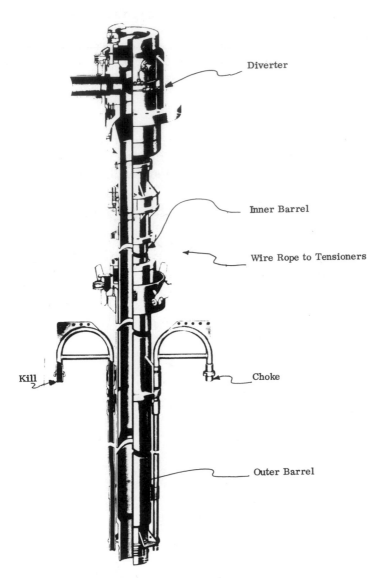

Fig. 2–15   *Telescoping joint. (Courtesy Regan Forge & Engineering Co.)*

function of allowing vessel heave and acting as a direct tensioning piston.

A diverter is located at the top of the telescoping joint. Depending upon the magnitude of the kick, the gasified mud is valved either onto the shale shakers or to port (Figure 2-16).

Tensioners

The inner barrel of the telescoping joint is connected to the gimbal under the drilling floor of a floating vessel (Figure 2-10). Wire rope runs through pulley systems on the deck down to the

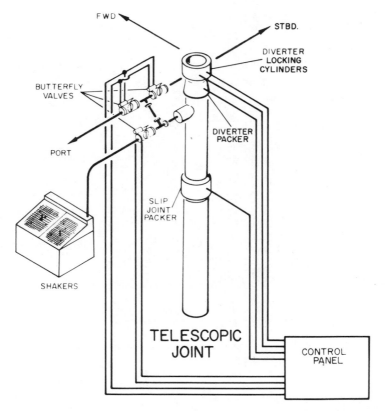

Fig. 2–16 *Telescoping joint schematic. (Courtesy Regan Forge & Engineering Co.)*

top end of the lower (outer) barrel of the telescoping joint. In the past, when floating drilling operations were restricted to shallow water, dead weights were connected to the deck end of the wire rope to maintain riser tension and prevent buckling collapse of the riser string.

Constant tensioners are now used having a total capacity ranging from 240-640 kips. The capacity of each individual tensioner is between 60 and 80 kips. The wire rope from the lower barrel of the telescoping joint is fed around pulleys on each end of a main hydraulic cylinder. Fluid in this cylinder is ducted to an air-over hydraulic bladder type of accumulator. The air end of the bladder accumulator is then ducted to a bank of air pressure vessels and the relatively constant tension is maintained by constant pressure with large changes of air volume due to the compressibility of the gas. A typical tension force-stroke curve is given in Figure 2-17. The riser tensioner main cylinders mounted on the mast substructure can be seen in Figure 2-11.

Total mechanical and hydraulic friction in any tensioning system is about± 15%. The variation from constant tension due to nonuniform pressure with varying volume depends upon the volume of the air pressure vessels used and is usually designed for 5-10%. The total variation from constant tension is then usually less than about 20-25%.

Riser tensioners are required for water depths greater than 250 ft. The critical buckling length, L_{cr} in feet, can be calculated by:

$$L_{cr} = 2.75 \left(\frac{EI}{w}\right)^{1/3}$$

where: E is the modulus of elasticity in psi
I is the moment of inertia in in.^{4}
w is the weight of the riser in water in lb./ft.

Riser Tension Guideline

W. L. Clark of Hydril has developed a useful procedure for sizing the guidelines for riser tensioners. His procedure is summarized as follows:

The riser tension guideline is a rough order of magnitude

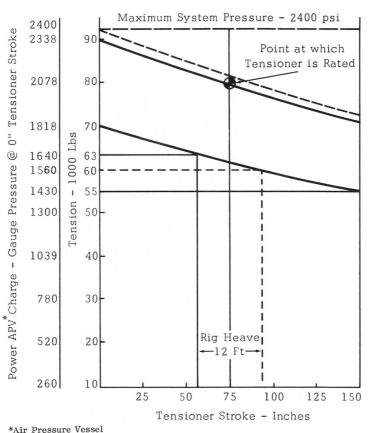

80,000 Lbs Tension
415 U.S. Gal Air Pressure Vessel – K = 1.1

*Air Pressure Vessel

Courtesy Rucker Control Systems

Fig. 2–17   *Tension versus stroke. (Courtesy Rucker Control Systems)*

calculation only. More complete and involved methods are required for complete accuracy.

This method of calculation is based on the weight of the riser pipe, the suspended pipe, and the mud column.

Nominal tension equals weight of the riser system in water plus 20% so that positive tension will be maintained at all times during heave motions.

Example:
> Weight of 20 inch OD riser x ½ inch wall on water $= -33$ lbs./ft.
> 18 lbs./gal. mud $= 264$ lbs./ft. *between ID of riser and OD of drill pipe*
> Nominal tension for 800 ft. of water, $T_{nom} = (800)\,(264\text{-}33)\,(1.20)$
> $$= 222 \text{ kips}$$

For purposes of safety, the tension system must be operated at a level high enough to provide the minimum tension level (that required to maintain the bottom joint of the riser pipe in positive tension) after losing one tensioner unit.

Thus:
$$\text{Operating Tension} = \frac{\text{Minimum Tension}}{\text{Safety Divisor}} = OT$$

$$\text{Minimum Tension} = \text{Riser System wt. plus 5\%}$$

$$\text{Safety Divisor} = \frac{\text{No. of Tension Units less one}}{\text{No. of Tension Units}} = \text{S.D.}$$

Therefore:
$$\text{S.D. for a four tensioner system} = \frac{4\text{-}1}{4} = 0.75$$

$$\text{S.D. for a six tensioner system} = \frac{6\text{-}1}{6} = 0.83$$

$$OT_4 = \frac{(800 \text{ ft.})\,(231 \text{ lbs./ft.})\,(1.05)}{0.75} = 259 \text{ kips}$$

$$OT_6 = \frac{(800 \text{ ft.})\,(231 \text{ lbs./ft.})\,(1.05)}{0.83} = 234 \text{ kips}$$

PRODUCTION RISERS

On fixed production platforms the tubes running from the platform deck to each subsea wellhead are called connectors.

After the oil runs up the connectors and thcough the processing equipment, it is then pumped from the deck down to the marine pipeline through the production riser. The three basic types of production risers are the J tube, the reverse J, and the bending shoe methods.

The production riser actually contains a bundle of pipes. These individual pipes include the production, solvent, and vent lines. The bundle may also include lines for gas lift, gas injection, water injection, air supply, and air return.

There are, of course, always exceptions to the conventional riser design mentioned above. One of these noteworthy exceptions is the phased operation in the Ekofisk field[7] in the North Sea. The initial oil discovery in the Ekofisk area was made in 1969 by the Phillips Norway Group with Phillips Petroleum as the operator. Development work began in November 1970 and production by sea loading began in the summer of 1971. This was the first development work in the middle North Sea, where the water depths go up to 230 feet.

This Phase I test production was brought on stream by converting the Gulftide exploratory jack-up into a production platform. In this case four wells were brought on stream with dual 4 inch pipelines running from each well to the Gulftide, connected to the caisson base, then up through a riser caisson located in the lattice jack-up leg.

DESIGN AND WATER DEPTH LIMITATIONS

Most floating exploratory vessels are now outfitted with a riser system for a maximum water depth of about 1,200 ft. Four or five new drillships in operation or under construction will have riser systems for the extreme water depth of 3,000 ft. A design analysis has been performed on one particular advanced riser system for 6,000 ft. water depth. From that study it appears that 6,000-8,000 ft. of water is the practical limit for the conventional riser system.

A drawback with risers for over 6,000 ft. water depths is that the static head of heavy drilling fluid (e.g. 20 lb. mud) could be about 7,500 psi at the mudline and proportionately greater with drilling depth compared to the normal water pressure of about 3,000 psi at the mudline. It is believed that this extraordinarily high pressure could cause severe fracturing of the formation. A possible solution might be to use gas injected mud in the drill string and artificial gas lift for the mud column in the annulus between the drill string and the riser with injection at the wellhead.

A shortcoming of this approach is that mud monitoring and control for blowouts would become even more difficult due to the problem of distinguishing between the injected gas and a kick.

The major problem with deep water riser systems for 2,000 - 6,000 ft. water depths is finding a practical balanced design for variable water depths. For deep water the riser string can become heavy enough from gravitational loading alone to severely stress the riser. The required tensioner sizes would then become enormous.

Augmented Riser Buoyancy

The maximum tension in the riser occurs at the top of the riser string and decreases with water depth. In very deep water, some type of added buoyancy is required to stay within the practical limits of the tensioner system. However, it must be remembered that the current drag force goes up as the square of the velocity and increases with riser diameter. This means that the outside diameter of the added buoyancy cylinders should be minimized. The optimum design riser has the smallest diameter and lowest possible wall thickness consistent with allowable stress levels and allowances for wear and/or abuse.

Plastic foam cylinders fastened to the riser have been used, and buoyant steel chambers have been suggested. The foam cylinders rapidly deteriorate, however, and require constant maintenance.

If a section of the riser system is made too buoyant, failure of a riser joint could hurl the riser up through the derrick floor like a missile. The requirement then for deep water riser systems is to make the riser neutrally or only slighly positively buoyant, with the buoyancy distributed over a long length of the riser.

Special Design Considerations

Sometimes it is desirable to make a first attempt at design with approximate methods, such as ignoring dynamic effects. When this is done, a safety factor of about 35% should be used on the riser design stress. This, of course, would not be satisfactory for large vessel motions in extreme environments which will be encountered in storms immediately before well abandonment. For example, in water depths of 500-1000 feet with typical riser designs, the dynamic stresses due to 36 ft. waves would be 8-10 ksi, not counting the dynamic stresses due to vessel motions.

Normally the outer surface of the riser should be smooth and with as few protuberances as possible. However, in exceptional cases special frames to preclude vortex shedding might be required. The water greatly dampens any vibrations due to resonant conditions, but special conditions where high currents could cause riser damage should always be investigated.

ANALYTICAL REVIEW OF
EXPLORATORY RISER SYSTEMS

Analyses for exploratory riser design must include a large number of variables. The combined stress in the riser is due to both axial and hoop or circumferential stress. Hoop stress is caused by internal pressure of the mud column and pressure with kicks. External forces on the riser result from hydrostatic pressure which is calculated for the given water depth.

The total axis stress is due to the tensioner load, riser buoyancy, riser weight, mud weight, current forces, and wave forces. With large surge or sway of the vessel the dynamic forces on the riser string must also be considered. A schematic of these stresses is shown in Figure 2-18.

Various procedures are used for analysis of the riser, depending upon the required objective. Simple catenary equations can be used to get a ballpark type of solution with results that may be off several hundred percent. This would be adequate for making rough comparisons of the effects of different water depths and diameters or wall thicknesses of the riser string.

For more accurate correlations, finite difference procedures

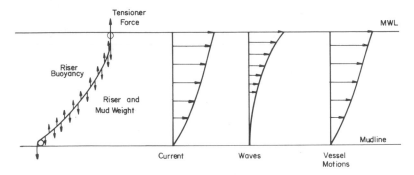

Fig. 2–18   *Forces on the riser string.*

are used to solve differential equations. The finite difference approach gives excellent results for a very simple type of riser. However, a separate differential equation is required for each change in cross section and the joints of different cross sections are tied together mathematically by compatibility conditions. The problem with the finite difference approach is that the procedure becomes totally unwieldy and impractical for analyzing a design with many variations in riser diameter, wall thickness, floatation material on the riser, or any other variables.

Conversely, the finite element approach for design and analysis finds considerable favor since this approach is virtually unencumbered by limitations or conditions. For example, the digital programming and computer usage is essentially the same for any number of variables, such as riser wall thickness, added buoyancy, etc. With the finite element approach, other variables can also be considered that cannot be done on a practical basis by any other method.

This means that any current distribution can be considered (Figure 2-19), hydrodynamic drag effects of the riser couplings and the choke and kill lines can be accounted for, loads due to any wave height can be considered, the influence of the telescoping joint can be fully analyzed, etc.

Fisher and Ludwig[2] have given generalized results that are very helpful for obtaining ballpark type answers. The value of their work is that the approximate response of any design can be quickly determined using their charts (Figures 2-20, 2-21).

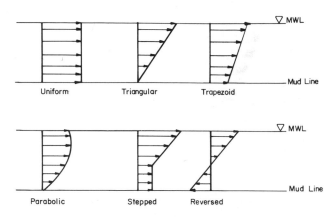

Fig. 2–19 *Common design current profiles.*

However, detailed and accurate analyses such as with a non-linear finite element routine must be used for either deep water conditions or severe environmental conditions.

The digital computer riser program, ETA/FLEXRIS, is a non-linear finite element program. Nonlinear means that the equations for displacement are second order rather than first. (Second order displacement functions are required because of the large displacements and relatively low stiffness of the riser cross section.) The nonlinear ETA/FLEXRIS program has all of the versatility of finite element programs. With ordinary standards of programming, the linear programs can be off by as much as 100%. The nonlinear programs give solutions with an accuracy of about 1%.

CONCLUSIONS

1. The technology of riser design and operations for 200-800 ft. water depths is well established. Special considerations are required for water depths greater than 800-1,500 ft.
2. Innovative designs are needed for 1,500-3,000 ft. water depths to reduce riser running time and to decrease the possibility of environmental damage due to extreme winds, waves, and vortex shedding.

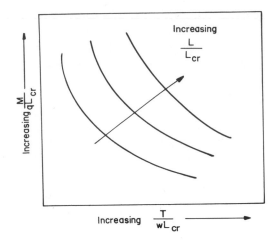

Fig. 2–20   *Maximum moment for vessel offset or uniform current.*

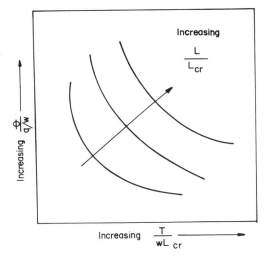

Fig. 2–21   *Rotation of bottom ball joint for vessel offset or uniform current.*

3. Because of the possibility of fracture of the geological formation with excessive mud pressure, radically different designs and procedures may be required for 3,000-6,000 ft. water depths.

4. Nonlinear finite element computer programs seem to be the best and most universal method for general design and analysis.

5. Computer simulated operations with various sets of wave and current forces and vessel motions can be used to set more realistic limits for drilling and hang-on conditions thereby decreasing weather related downtime.

REFERENCES

1. Butler, H. L., Delfosse, C., Galef, A., and Thorn, B. J., "Numerical Analysis of a Beam Under Tension," *J. Struct. Div., ASCE*, Oct. 1967, pp. 165-174.

2. Fisher, W., and Ludwig, M., "Design of Floating Vessel Drilling Riser," *J. Petr. Tech.*, March 1966, pp. 272-280.

3. Harris, L. M., *Deepwater Floating Drilling Operations*, Tulsa: Petroleum Publishing Co., 1972.

4. Kennedy, J. L., "High-Angle Holes, Supply Hurdles Complicate North Sea Drilling Work," *OGJ*, June 24, 1974, p. 132.

5. Morgan, G.W., "Riser Design Criteria and Considerations," *Petr. Engr.*, Nov. 1974, pp. 68-74.

6. Morgan, G. W. "General Aspects of Riser Design and Analysis Procedures," *Petr. Engr.*, Oct. 1974, pp. 36-48.

7. Wilson, R. O., and Martin, M. R., "Deepwater Pipelay for Central North Sea," OTC Paper 1855, 1973.

8

Surface Motion Compensation

William L. Clark
Hydril Company

Wireline Motion Compensation-Logging

A motion arrestor for nullifying vessel heave in wireline operations, especially in well logging operations, is available. This compensation device hangs below the hook and uses a working wireline reeved from the top of the riser pipe around the motion arrester working sheave, and to the drill floor. The logging sheave and working sheave are connected to the same yoke, which is suspended from the motion arrester by a pneumatic spring (special tensioner).

As the rig heaves upward, the working sheave and the logging line are retained in a relatively fixed place in space (as controlled by the two-part line). As the rig heaves downward, the sheaves again remain in a relative place in space as controlled by the pneumatic spring. This compensation motion has the effect of nullifying relative motion between the rig and earth generated by heave of the vessel. This eliminates the rig motion that would otherwise be superimposed by the data read by the logging sonde.

This compensation device is not required when a drill string compensator (DSC) is installed. The DSC has the capability of performing this function in addition to many others.

Drill String Motion Compensation

The major application for a drill string compensator is to nullify rig heave that would be imposed on the drill string. This motion nullification significantly improves the operation of the following procedures:

a. *Drilling*

The DSC maintains a virtually constant bit weight, unaffected by rig heave. It improves penetration rates and significantly improves drill bit life. The DSC permits instant and easy drill bit weight changes at the surface, without tripping the drill string, thus eliminating many trips of the drill string.

b. *Landing BOP Stack*

The DSC permits a relatively soft landing of the BOP stack and/or marine riser, not only in a safer manner, but in rougher sea conditions or higher rig heave conditions than would otherwise be possible.

c. *Landing Casing*

The DSC permits safe landing of casing in sea conditions and/or rig heave conditions that would otherwise be impossible or unsafe.

d. *Safety Control*

The DSC can eliminate the motion of the drill string in the BOP stack. This eliminates wear of BOP seals due to rig heave and drill string motion incurred when closing the rams or annular BOP's on the drill pipe.

e. *Miscellaneous Operations*

Operations that would otherwise be hampered or impossible with the drill string moving with the rig heave are now possible because the DSC eliminates drill string motion during many operations.

Basic Methods of Drill String Compensators

All present drilling rig DSC's are an air spring tensioning device. They are passive devices that function based on the difference in the suspended weight of the drill string and the tension level set in the DSC.

It is possible to manufacture an active-servo controlled design, but at this time only one or two such units have been manufactured. They have been used on coring vessels. The following discussions cover the popular passive systems which are predominantly in use.

The weight on the drill bit equals the weight of the drill string less the DSC tension setting. Tension levels are controlled in the DSC exactly the same as in a riser tensioner. The basic techniques and technology developed for riser tensioners were applied to drill string compensators.

During drilling, the drill string weight is supported by the hydraulic-pneumatic cylinder of the DSC and the drill bit weight on the bottom. The cylinders are interconnected to the air pressure vessels (the same as the riser tensioners). Control of the air pressure in the air pressure vessels determines the tension level. Proper DSC drilling techniques always require a DSC tension setting to be less than the weight of the drill string.

As the rig heaves upward, the support cylinders must stroke to extend the DSC and thus to compress air from the cylinders into the air pressure vessels. The large volume of the air pressure vessels controls the variations in pressure due to air compression, the same as in the riser tensioners. This cylinder stroking maintains the preselected support load (tension) and thus maintains virtually the same weight on the drill bit.

As the rig heaves downward the support cylinders retract the DSC. Downward heave tends to place more weight on the drill bit; but as the cylinders are maintained at the preselected air pressure level, the cylinder retracts the DSC to maintain the preselected load (tension) and thus maintains the drill bit weight. During retraction of the DSC, air expands from the air pressure vessels to the cylinders maintaining the desired pressure level.

Ratio of the cylinder volume and air pressure vessel volume is important. The volume selections by each manufacturer optimize the equipment required versus the benefit derived from large volume air pressure vessels. An explanation of these effects can be seen in the section on Pressure Control in Cylinders in the "Tensioning Systems" section.

The DSC application requires closer control than the riser tensioners. The load variation of the DSC, as related to heave stroking or compensation stroking, is a percentage of the total supported weight. However, the desired control is the variation in drill bit weight, which is a fraction of the drill string weight.

For example assume:

a. Drill string weight of 250,000 lbs.

b. Desired drill bit weight of 50,000 lbs.

c. A DSC variation of ± 6%

The stroking variation equals ± 6% of 200,000 lbs. or ± 12,000 lbs., which is 24% of the desired drill bit weight.

The example above also illustrates the importance of installing the pneumatic and hydraulic system in accordance with the manufacturer's specifications. Pressure drop due to air/oil flow through the piping system causes a direct degradation of performance. System designers have optimized their system requirements and the applicable specifications should be followed.

Basic Types of DSC's

a. *Deadline*

A tensioner can be mounted in the deadline of the drawworks system to control the weight on the bit. This technique has been used on one rig and functions well as a safety device, but was found to be too mechanically inefficient for drilling operations.

b. *Crown Block*

A tensioning device supports the crown block and thus the drill string. By supporting the crown block with controlled tension the compensator becomes a motion nullification device by raising and lowering the crown block. This then raises and lowers the traveling block and hook to nullify motion or to isolate rig motion from the drill string. This technique is presently in operation on several drill rigs and has proven to be effective. The principles of operation and control used in this design are similar to the traveling block DSC.

c. *Traveling Block DSC*

A tensioner device to support the drill string is connected between the traveling block and the hook to become a motion nullification device. Its tension level is controlled by techniques identical to a riser tensioner.

Techniques of Traveling Block Drill String Compensator

a. *Tension Type Cylinder*

This technique applies to a cylinder(s) with high pressure on the rod side of the cylinder(s) between the traveling block and the hook (Figure 2-22). In order to provide lubrication to the cylinder an air/oil accumulator is used on the rig floor or on the derrick (mast) and oil flows through the hose loop to the cylinder. Pressure in the cylinder controls the DSC tension level in the same manner as in a riser tensioner. The weight on the drill bit is determined by the weight of the drill string minus the tension setting of the DSC.

In the event of a drill string breakage the oil flow input to the cylinder must be instantly shut off to prevent damage to the cylinder, surrounding equipment, and personnel.

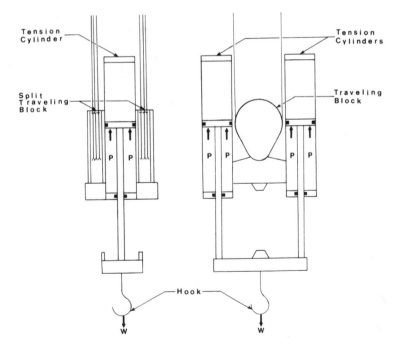

Fig. 2–22 *Drill string compensators–tension type.*

b. *Compression Type Cylinder*

This technique applies a cylinder with high pressure air on the blind side of a cylinder between the traveling block and the hook (Figure 2-23). Lubrication and safety control are achieved by the low pressure air/oil reservoir on the rod side of the cylinder. Chain reeved around the cylinder in two parts of line delivers a compensation stroke twice that of the cylinder stroke. In other words, a 9 foot cylinder stroke delivers 18 feet of compensation motion. The use of a low pressure air/oil reservoir permits use of high pressure air directly on the blind side of the cylinder, so that air is flowed through the stand

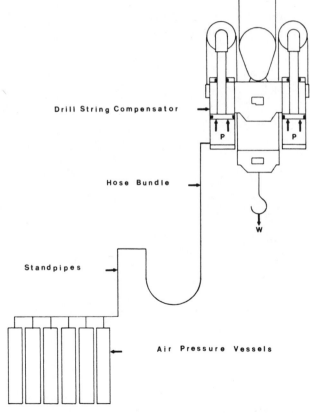

Fig. 2–23 *Drill string compensator–compression type.*

pipe and hose loop for a reduced pressure drop and reduction of shock loading.

Techniques of a Crown Block DSC

Compression loaded cylinders (Figure 2-24) support the crown block. The crown block is mounted on a framework-supported dolly atop the derrick, which allows vertical motion of the crown block. Controlling the pressure in the support cylinders determines the tension level or load support level of the crown block.

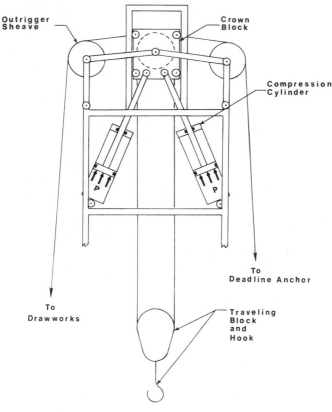

Fig. 2–24 *Crown block compensator.*

The weight on the drill bit equals the weight of the drill string minus the tension setting of the DSC.

The crown block DSC moves the compensation device to the crown block rather than locating at the traveling block. This location eliminates the hose loop required by a traveling block DSC, but present crown block DSC designs are significantly heavier.

Articulated out-rigger sheaves are mounted on the crown block support mechanism. These sheaves are connected to nullify motion of the traveling block which would otherwise be generated by raising and lowering of the crown block relative to the drawworks. This causes motion of the drawworks wireline system and would otherwise be seen as motion in the traveling block.

Operation of a Drill String Compensator

a. *Tension Level*

The tension level of a DSC is determined by the air pressure in the cylinder or the oil pressure controlled by the high pressure air in the accumulator.

Bit weight = drill string weight minus DSC tension

When a DSC is in normal operation, the large volume air pressure vessels are directly connected through piping, the stand pipe, and a hose loop to the DSC cylinders (with an air/oil interface on the tension type cylinders). The tension level is increased by compressing more air into the closed air system (air pressure vessels). It is reduced by venting air from the air pressure vessels into the atmosphere. Control of the air pressure in the air pressure vessels and DSC system is identical to techniques utilized in the riser tensioner systems.

b. *Commence Drilling*

After setting the desired tension level in the DSC and making the drill string connections, lower the drill string into the hole until the drill bit contacts the hole bottom. At contact, continue to lower the traveling block to permit the DSC to stroke about its midpoint. Note that the tension level of the DSC is always less than the total weight of the drill string. Therefore, when picking the drill string up from the slips, the compensator will extend its full length before the drill string is lifted from the slips.

When reaching the bottom of the hole and landing the drill bit, the compensator will begin to retract and support weight of the drill string equal to its tension level. This leaves the difference in the weight and DSC tension on the drill bit as previously calculated. Lowering the traveling block to approximately half the stroke of the DSC will set the DSC to operate about its mid-stroke. As the bit drills off, the driller continues to incrementally lower the traveling block to maintain the DSC stroking about its midpoint. As long as the DSC is operating within its stroking limits, the weight on the bit will be controlled by the DSC and will be maintained at its preselected load.

c. *BOP Stack Landing*

A relatively soft landing of the heavy BOP stack can be achieved utilizing the DSC. One common technique that has been successfully used is to set the tension level of the DSC a few hundred pounds less than the weight of the package to be landed on the ocean bottom. When beginning the initial lift from the drill rig (slips or spider), raise the traveling block to extend the DSC to its complete length (the weight exceeds the tension setting). When reaching the ocean floor with the package, the initial contact on the ocean bottom with the package will allow the compensator to begin retracting in its stroke.

However, the DSC will continue to support most or almost all of the load (an amount equal to its tension setting), leaving the ocean bottom structure supporting only the difference in the tension setting and the weight of the package, which can be adjusted to very light loads as compared to the total weight of the BOP stack. This technique permits landing and retrieving the BOP stack in much rougher sea conditions, and in a safer manner, than would otherwise be possible.

d. *Casing Landing*

Landing casing is accomplished in the same manner as described in the BOP stack landing. This relatively soft landing of casing permits the operation to take place in much higher rig heave conditions without damage to hangers and hanger seals than could be achieved without such a device.

Summary

The DSC is a tool which permits more efficient drilling operations in normal sea conditions, and in rougher sea conditions it permits drilling to continue when they might otherwise be impossible. A DSC saves rig time, and thus money, on each location.

Time on location is reduced. The amount of time reduced or saved varies between locations due to area conditions, rig design, weather conditions, etc. However, use of the DSC can always reduce time. Drilling of the hole can be accomplished faster with less days, fewer trips, and fewer drill bits. Setup and relocation times are reduced and can also be accomplished in rougher sea conditions. Thus, the total time for a single hole completion is reduced, sometimes saving as much money on one location as the entire cost of the DSC system.

Drill string compensator (400,000 lb. unit) traveling block type. (Courtesy of Rucker Control Systems)

Drill string compensator (400,000 lb. unit) traveling block type. (Courtesy of Rucker Control Systems)

Drill string compensator (400,000 lb. unit) traveling block type. (Courtesy of Vetco)

Part III

Blowout Prevention

9

Offshore Blowouts and Fires

Dr. M. D. Reifel

Marathon Oil Company

INTRODUCTION

Blowouts are of two types, surface blowouts and subsurface blowouts. Surface blowouts are possibly accompanied by fire, explosion, pollution, third party damage, and property damage to the drilling rig and platform. They also expose the company involved to potential unsatisfactory public opinion. If the blowout is uncontrolled, a valuable reservoir may be needlessly depleted. On the other hand, a subsurface blowout which leaves no impression on the surface can be equally devastating to a hydrocarbon reservoir. Although less spectacular, the subsurface blowout can be more difficult to control.

Controlled blowouts may be defined as the flow of hydrocarbons from the wellbore which is contained by the rig equipment. A catastrophic blowout is generally one that results from a failure of rig equipment/personnel, from natural causes such as a storm, or from failure of mooring equipment in the case of a floater. Catastrophic blowouts result in uncontrollable flow from the wellbore.

This section is concerned with catastrophic surface blowouts. Some technical aspects of well control equipment used to control blowouts are discussed in the section on blowout prevention.

INDICATIONS OF BLOWOUTS

Drilling is usually accomplished with the imposition of a positive hydrostatic head larger than the formation pressure. A number of situations may be encountered which can alter this balance, as when the drilling mud weight is too low and formation fluids can flow into the wellbore. An abnormal pore pressure gradient, greater than the pressure of the mud being circulated, would likewise result in an upward flow. Drilling operations involving the raising and lowering of packed drill strings and other near gauge equipment out of the well could swab the well in. Similarly, running large gauge hookups into the well could cause a fracture of the well, resulting in lost circulation and subsequent flow. Indications of potential blowouts at the surface vary from slightly gas cut mud to heavily gas cut mud to production of gas and mud.

Kicks

A kick results from an influx of fluids from the formation into the wellbore. Kicks derive their name from the flow behavior observed as formation fluids approach the surface. The trouble starts when these fluids—oil, water, or gas—enter the wellbore. Formation fluid influx normally creates a pit gain and well control becomes increasingly more difficult as the pit gain increases.

A small influx of gas into the bottom of the wellbore will occupy a large volume near the surface. Also, the rate of increase in volume accelerates constantly as the gas nears the surface. A large influx of gas can fill the annulus as the gas expands. Since the density of gas is low, the hydrostatic pressure exerted by the gas column is low, and excessively high back pressure on the casing would be required to balance formation pressure. This is why it is so important to shut a kicking well in with a minimum amount of gain in the pit.

A gas influx must be allowed to expand as it rises. If the gas is allowed to rise without expansion, the gas reaches the surface at the same pressure at which it originally entered the wellbore. The casing and other surface equipment are subjected to a pressure equal to the formation pressure. This pressure, exerted above the hydrostatic mud column, is transmitted through the mud column, resulting in about a twofold increase in pressure exerted on the bottom of the hole.

Thus, lost circulation, underground blowouts, and blowouts at the surface too often result from attempting to circulate without permitting an influx of gas to expand. Allowing gas to rise in a controlled expansion keeps bottom hole pressure constant, reduces pressure gradients at the casing shoe, and reduces strain on surface equipment.

A kick while drilling is indicated by an increase in pit level and may be due to:

1. Very permeable formation, unbalanced by mud pressure;
2. Very permeable formation, slightly unbalanced by mud pressure;
3. Low permeability formation, unbalanced by mud pressure;
4. Kicks following a trip.

There is a tendency to forget the mud pit if the hole is filled and checked after the pipe is pulled. Because of the possible presence of a gas slug, flow can begin after the pipe has been pulled and enough time has elapsed for the slug to move up the hole. Busy crews sometimes forget to observe pit level, flowline, and hole. Before enough mud flows to show on a pit level recorder, a small flow will begin at the flowline. If this small flow is noticed, there will be time to take control measures before a serious problem develops.

Gas Cutting

Even though fluids from a formation surrounding the bore hole cannot intrude into the well when hydrostatic pressure exerted by the mud column is greater than formation pressure, fluids contained in the pore spaces of formation materials removed by the bit can cause gas cutting. As the bore hole is deepened by continued drilling, solid particles of the formation must enter the mud circulating system as drill cuttings. Any oil, gas, or water not held in the drill cuttings also enters the mud system as soon as the formation is drilled up. Even if the formation is not permeable, fluids are released from pore spaces on the

242   Reifel

surface of the cutting. Changes in the mud system *in* are a reflection of the contents of bottom hole sediments and their pressures.

As the cuttings move up the annulus, hydrostatic mud pressure is constantly reduced and more fluids can be released from the cuttings. Regardless of the mud weight, hydrostatic pressure in the bore hole decreases to three fourths of the bottom hole pressure at three fourths of drilling depth, to one half at one half of the depth, and so on until zero pressure is reached at the surface. No amount of mud weight can prevent fluids in drill cuttings from entering the mud system.

Gas-cut mud does not necessarily indicate a need for higher mud weight. Since most of the gas expansion takes place in the upper part of the hole, very extreme gas cutting reduces bottom hole hydrostatic pressure by a small amount. Lost circulation usually results from needlessly increasing mud weight in a futile attempt to control gas from drilled formation solids. However, conditions can be such that a mud weight increase is advisable.

Mud Equipment

Danger of a blowout from an unobserved kick increases with time. Forewarning and indication of trouble are based on changes between the mud system *in* and the mud system *out*. The mud system *in* is predetermined for formation pressure containment and penetration rate control.

Forewarning devices, developed for continuous and/or intermittent surveillance of mud systems, can detect small changes before they become visually apparent. Since kick control is so sensitive to the volume of pit gain, some device other than a nut hanging on soft line is required for monitoring pit levels. Among others, the Warren "Barrel-O-Graf" or the Martin-Decker "Mud Volume Totalizer" both indicate and record the total barrels in the active mud system.

Although small changes in pit level are usually detectable with these devices, an earlier warning of flow change can be provided by the Warren "Flo-Sho" or the Martin-Decker "Mud Flow-Fill," both of which indicate and record the flow rate in the mud flowline during hole filling, operating, and drilling.

Changing mud weight and temperature between "mud in" and "mud out" can be measured by devices such as the Martin-Decker "MDT" or the Warren "Mud-O-Graf." These sensing and recording devices are usually employed as complementary components for monitoring mud volumes, flow rates, and density/temperature.

Although a considerable portion of gas will break out of drilling mud at the shale shakers, a vacuum mud degasser is needed to more completely remove gas. A degasser is still needed when the rig is equipped with a blow-box or mud-gas separator. When kicks are anticipated, the degasser should be kept in operation. Gas-cut mud should be weighed at the degasser discharge for pressure control. Mud weight coming out of the hole should also be monitored ahead of the separator and/or degasser.

A rate of penetration recorder is essential for detecting a pressure transition zone. The type marketed by the Bell Corporation aids interpretation since the plotted curve also indicates to some extent the nature of the formations being drilled. Often an excellent correlation can be made between the SP log and the rate of penetration plot.

CAUSES OF BLOWOUTS

Good planning, execution, and analysis will reduce frequency and severity of kicks, while a true commitment to understanding kicks will eliminate loss of wells resulting from blowouts and stuck pipe. Nevertheless, kicks can occur in any drilling operation conducted with a view towards attaining maximum effectiveness. These relatively infrequent kicks need not be unduly dangerous nor time-consuming when proper and prompt action is taken.

Pressure unbalance leading to a kick may result from one or more of the following causes:

1. Poor well planning;
2. Failure to keep the hole full;
3. Swabbing;
4. Lost circulation;
5. Mud weight too low.

A blowout is an uncontrolled kick. Kicks develop into blow-outs for one or more of the following reasons:

1. Lack of early detection;
2. Failure to take proper initial action;
3. Lack of adequate casing and/or control equipment;
4. Malfunction of control equipment.

Equipment failure is less of a hazard today than it has been in the past because the equipment is more reliable. Better well control practices also result in the equipment being subjected to less strain and fatigue. Perhaps the industry has also learned to maintain blowout equipment better. On the other hand, wells are becoming deeper and over-pressures in many areas are becoming the rule rather than the exception. On balance, therefore, we can allow no compromise in keeping oil well control equipment in first class condition.

Most blowouts occur not because of faulty equipment but as a result of human error, poor practices allowed to multiply until a mistake is made, or a combination of mistakes which bring the well into a critical condition. It is only after this stage has been reached that so much depends on the design and condition of the blowout preventer stack.

Many blowouts are initiated at about 2:00 or 3:00 in the morning, or during some holiday season when precautionary routine is curtailed or even eliminated. The key to blowout prevention usually lies not in spending large sums of money on high pressure preventer equipment, but in teaching crews to identify foreign fluid entry into the wellbore at the very earliest time and then taking prompt action to remove this fluid from the well. The problem increases in size and complexity in proportion to the amount of foreign fluid allowed to enter, since greater surface pressures are thereby required to replace the head of drilling fluid unloaded from the well.

Whatever the cause, each situation must be met with an intelligent approach that will lead to the logical control of the well. A prepared plan of action for foreseeable eventualities should be available. This plan, along with the necessary equipment, must be ready before the blowout occurs. Blowout prevention is a frame of mind of the drilling crew and supervisory staff. The determination of operators and management to eliminate blowouts is far more valuable than any equipment used to control blowouts.

EMERGENCY PROCEDURES—KICKS

A drilling break, sloughing formation, prolonged severe gas cutting, or suspected increase in mud flow at the flowline may give an early indication of a well which will flow. Any of these early signs should be checked so that the volume gained on the kick will be limited to a small volume. Any indicated rise in pit level must be followed by emergency procedures of closing in the well and restoring conditions to normal. No time can be wasted, as the more foreign fluid that enters, the more difficult and serious the situation becomes. Moreover, the closed-in condition of the well must also be regarded as a temporary condition, required only to check the degree of lost control. It is a prelude to the next step of circulating the well over a choke to regain normal control.

Shutting in should consist of closing down the pump, raising the kelly, and closing the annular preventer, thus diverting the flow through the choke manifold after which the flowline valve is closed to completely shut off flow from the well. This stepwise procedure is designed to prevent shock loading of the elements of the blowout preventer stack. Total elapsed time from recognition of the emergency condition to complete well closure should not be more than one minute.

The well must also be closed in only long enough to check pressures at the wellhead. Readings of pressure on standpipe and annulus should be taken as soon as the well is closed in completely. Any attempt to leave the well closed in may lead to complications because of the tendency of gas to rise in the annulus and raise wellbore pressures through pressure inversion. If for any reason the well has to be kept closed any longer than required for pressure reading, the gauges must be watched closely for rise in pressure indicating that pressure inversion is in progress.

If necessary, the annulus must be bled down frequently to avoid rising wellbore pressures from breaking down the formation, with danger of underground flow between formations or even to the surface behind the casing cementation. Bleeding down should be stopped as soon as standpipe pressure fails to drop further, indicating fresh flow is occurring in the wellbore. When circulating, gas does not enter the drill pipe. But when shut down with the well closed in, gas can and does enter the drill pipe causing pressure inversion and sometimes formation fracture.

Only under extreme circumstances, therefore, should circulation be delayed after taking shut-in pressures. Close-in time should not exceed two to three minutes, even if drill pipe pressure is still rising slowly.

If the drill pipe is off bottom at the time of flow, it should be returned to bottom, stripping it in if necessary. Returning to bottom allows the kick to be controlled with a smaller weight increase. When running back to bottom, a volume of mud equal to the pipe displacement (but no more) should be allowed to flow from the well.

As soon as possible the killing operation should be started by circulating the well over a choke and, if necessary, increasing the density of the drilling fluid. A kick should be circulated out before the mud system is weighted up so that the drilling fluid is kept moving, and the formation fluids are moved from the drilling collars. The well must be circulated on the choke to prevent breaking the formation down while the mud system is weighted up. After the drill pipe pressure and casing pressure are stabilized at the correct values, and the mud pump is running at the required rate, the choke should be controlled to maintain drill pipe pressure according to the pumping schedule.

EMERGENCY PROCEDURES-BLOWOUTS AND FIRES

In the case of an oil or gas well blowout, action should be designed to protect human life and control the disaster as rapidly as possible. Very often these catastrophic blowouts are accompanied by fire because ignition sources abound around a drilling floor. Secondary fire problems are caused by the use of ordinary combustibles on drilling rigs and/or platforms.

If a drilling rig is on location, all engines must be shut down and all feeder lines into the installation must be shut in. All electric power lines in the area should also be cut off. All personnel must be accounted for and evacuated to a safe distance (off of the rig or platform). In the event of an injury, medical and/or ambulance service must be called out. These services should be alerted in any event.

A small crew can then be used to clear the location of all equipment that can be safely moved. If liquid hydrocarbons or gas are in storage or in a gas plant on the rig or platform, these should be pumped or flowed from the affected installation when practical. The hydrocarbons should be removed from the installations as long as possible, with the fluid temperature being continually checked.

If liquid hydrocarbons are flowing or spilling in the offshore waters, the equipment and personnel necessary to contain and clean up these fluids should be activated. The appropriate fire fighting equipment available should immediately be used to fight the particular type* of fire, while service companies with fire fighting equipment are mobilized.

After initial procedures are implemented, appropriate company officials, governmental regulatory agencies, working interest owners, and contractor's management should be notified. All questions from the press or other parties concerning the situation should be routed through one man.

If the blowing well(s) does not bridge over, well control specialists should be mobilized. During the mobilization period necessary safety equipment (such as air tanks and masks in case of sour gas, brass tools, etc.), additional supplies of fire fighting chemicals, necessary mud materials, mixing and pumping equipment, and special well control tools should be ordered.

---

*Fires on rigs and platforms are classified into three categories:
   Class A — fires involving ordinary combustibles, such as wood, paper materials;
   Class B — fires involving inflammable liquids and/or gasses;
   Class C — electrical fires.

TECHNIQUES TO CONTROL
CATASTROPHIC BLOWOUTS

Control over a catastrophic blowout can be gained by any technique that blocks the escaping reservoir fluid either in the wellbore or in the formation. The method most frequently used is wellbore blockage, i.e., the recapping of a wild well. Relief well techniques, although more expensive to implement, have popular support because water pollution can be minimized by allowing the flowing wells to burn.

Relief Well Technique

Relief wells are sometimes drilled to establish direct connection with the wild-well bore hole. Then heavy mud is injected at a rate greater than the lifting capacity of the blowing well, until the well is brought under control. If the bottom hole pressure of the reservoir exceeds a pressure that can be balanced by feasible mud injection rates, the relief well may be deliberately aimed off the wild well landing point so that mud can be injected into the reservoir to plug the passageways open to the escaping reservoir fluids.

After a wild-well is brought under control with a relief well, final plugging operations are still necessary to gain permanent control over the well. These surface operations involve clearing away rig and/or platform wreckage around the wells, setting temporary wellheads, and killing the wells from the surface. With control of the wells assured, cleanup can continue by removing all of the wreckage before the platform and wells are completely restored. Usually the wells are restored by installing new conductor pipe and relanding the casing strings. Of course, some wells may be cemented in and abandoned.

One of the industry's most complex and precisely engineered directional drilling programs was successfully executed to control eleven wild wells which blew out on Shell Oil Company's Platform B in Bay Marchand, offshore Louisiana, in December 1970. A total of five rigs drilled 126,925 feet of hole[1] in ten relief wells to control the blowouts with 17 lbs./gal. mud. One relief well was used to kill two wild wells. It took 136 days to bring the wells under control at a total cost of about $28 million, including pollution control measures. Several million dollars of oil and gas was also burned.[2] The cost of restoration operations was estimated at $2.8 million in addition to the $1.8 million for a new platform (the old platform was lost at $5.2 million).

Four drilling rigs took 47 days to drill four relief wells to Amoco Production Company's wild wells that blew out on Platform B in Eugene Island, offshore Louisiana, in October 1971. It cost about $9 million to drill the relief wells, fight fires, and control pollution.[3] The cost of restoration was estimated at $5.4 million.

Explosive Snuffing Technique

In order to recap a burning wild well from the surface, the fire must first be extinguished. With the explosive snuffing technique the blast from a dynamite capsule positioned over the fire uses the oxygen necessary for conflagration and the fire is snuffed out.

As seen in Figure 3-1, the burning wells and rig or platform must be sprayed with a battery of jet nozzles to cool the metal so the fire will not reignite after it is snuffed (also to prevent deterioration of the wellheads and structure). Jet barges, equipped with fire-fighting pumps, hoses and nozzles, and capable of pumping up to 15,000 gpm of seawater on the wells and structure, are available in some drilling areas. A work platform (Figure 3-1) is usually set to provide spray capability and a stable work surface for the positioning of the explosive charge and for subsequent surface operations.

After the fire is extinguished, the spewing wellheads and conductor pipes must be cut clean with shaped charges so the wild wells can be capped. A long casing cap is lowered over the spewing conductors to direct the flow of oil and gas, and water is injected through the casing cap while the shaped charges are ignited to minimize the threat of reflash of the fire. Temporary wellheads are then installed on the conductor pipes to bring surface control to the wild wells.

The explosive snuffing technique has a major disadvantage in the potential oil spill with resulting offshore pollution. The fire

Fig. 3–1   *Chevron's platform C on fire in Main Pass, Block 41, offshore Louisiana.*

shown in Figure 3-1 burned for four weeks while Chevron Oil Company amassed the largest collection of pollution control equipment ever marshalled for offshore purposes, as they prepared to snuff the blaze and cap the flowing wells.[4] [5]

A crew of wild well control specialists was used to cap six wells after the fire was snuffed out with explosives on Chevron's Platform C in Main Pass, offshore Louisiana, in February 1970. Two additional wells were brought under control by injecting mud and cement into relief wells. Because the wells were allowed to flow for three weeks after the fire was snuffed out,[4] oil pollution damaged wildlife and two islands, drawing close public attention to the catastrophe, and causing postponement of an offshore lease sale. Louisiana oyster fishermen filed a $31.5 million federal pollution suit against Chevron, and the shrimp fishermen filed a $70 million suit.[5,6] Chevron was fined $1 million for the oil spill by a U.S. District Court in the first prosecution under the 1953 Outer Continental Shelf Lands Act.[7]

Diver Activated Well Killing Techniques

A well killing technique has been developed by Tenneco Oil Company and Brown Oil Tools, Inc.[8] Rather than fighting a fire at the surface of the sea where both wind and waves add to the difficulty and danger of the task, crews approach the well beneath the water surface where there is no danger of fire or explosion. In this relatively safe environment, divers can cut into the casing strings of a wild well, and then tap into and plug off the tubing that sustains the fire. This technique has the advantage of allowing oil and gas produced from wild wells to burn, minimizing offshore pollution during the killing operations.

Before the well plugging method can be utilized, access to the tubing strings that are blowing must be effected. This involves two general operations: (1) access through the platform/well debris to the well casing; (2) access through the casing program to the tubing strings. After diver reconnaissance of the general condition of the platform structure and wellheads, the decision can be made to begin access operations.

To gain access to the well casing does not require full scale salvage of the platform and wellheads (which may be done after the wells are brought under control). Instead, the divers must clear only sufficient space to allow travel to and from the well casings for the diver and equipment and provide work room around the well casings of interest. Temperature and sonic measurements on the well casings can be used to locate the casing strings containing blowing wells.

To provide working space, much of the tangle of debris can be moved aside with air tuggers and winches controlled from a work barge. If debris cannot be moved aside to provide access to the well casings, standard explosive devices can be used to sever the obstructing members. Straight, circular, and structural beam explosive-shaped charge cutters are widely used to produce slot-type cuts to any desired length, depth, and configuration.

Once access to the well casings is obtained the divers can begin access operations to the tubing strings. For cutting windows in casing a "picture-frame" cutter was developed.[8] It uses a combination of linear and circular shaped charges, and the cutter is designed so there will be no damage to a casing or tubing string inside the casing being cut even though the inside string may be laying against the wall in the line of the cut. Windows are cut in the successive casing strings until access to the tubing strings is obtained.

Before proceeding with the window cutting operations on a casing string (and almost certainly on the production casing string) the divers may hot-tap the annulus to check for pressure if communication with the well is suspected or is remotely possible. If the annular area contains pressure, this must be bled off or killed with mud or cement. It if is not possible to kill the annulus, the well must be shut in with alternate techniques.

Where water depth permits, a window is cut in the casing strings near the sea floor, and another window is cut at diver height above this window. If the water is not deep enough to allow two windows or if it is more expedient, one elongated opening may be used as seen in Figure 3-2.

A hot-tap device is attached to the tubing in the lower window, and an entry hole is cut into the string or strings which are out of control. In the upper window a crimping device is applied to the tubing to form a stricture (Figure 3-2). Sealer balls are

Fig. 3–2 *Hot-tap, injection canister, and crimping tool is shown installed on tubing string.*

introduced into the flow stream from a canister attached to the hot-tap device. The fluid source for injection is a flexible hose leading to the surface. The flow inside the tubing lifts the sealer balls until they are trapped by the crimp. The sealer balls are followed by a finer lost circulation material until an impermeable plug has been formed at the crimp, as shown in Figure 3-3.

Prior to and during the plugging operations, the flowing oil and gas burns instead of spilling in the water. With the plug in place, flow to the surface terminates, the fire dies, and the well is ready to be killed by pumping mud and/or cement through the hot-tap hole.

Hurtsteel Products, Ltd. of Canada has proposed a similar diver activated well killing technique called the SNUFF System.[9] With this system access to the inner casing string and/or tubing strings is obtained by the techniques described above. However, the wild well is plugged by injecting CO_2 into the tubing, or casing annulus, to form an ice plug until the well is brought under control. The wellhead equipment could also be frozen with CO_2 injections.

Fig. 3–3 *Plugged pipe section is milled open to show impermeable plug.*

Neither diver activated well killing technique has been used to kill an actual wild well. However, the Tenneco/Brown technique has been completely developed and successfully tested on a simulated blowout at an offshore location.[10] This technique is ready for commercial application with a complete set of the required tools and equipment assembled for emergency use.

Obviously, the diver activated well killing techniques cannot be used to control every catastrophic blowout (as mentioned above, the casing annuli must be killed). If debris from the burning wells and rig/platform structure are falling, the decision may be made to delay access operations until the situation stabilizes, construct a structural umbrella to protect the divers, or use alternate techniques to control the blowout. It should be emphasized that divers cannot be paid enough to commit suicide, so they won't go under a burning structure if the situation is one of high risk. Also, equipment and structural members do not fall instantaneously—the diving master has time to warn the divers to safety.

Figure 3-4 shows a situation suitable to diver activated tech-

Fig. 3-4   *Gas well blowout offshore Libya in 85 ft. of water.*

niques. This blowout occurred offshore Libya in 85 feet of water in August 1966. When the rig blowout control equipment failed during a kick, the drillship was moved off location ripping the riser, kill and choke lines, and hose bundle from the wellhead.

The blowout and fire shown in Figure 3-5 occurred offshore Australia in August 1969. The semisubmersible rig, drilling in 330 feet of water, could not be pulled off location after a tool joint ripped the packing element of the annular preventer during heave following a kick. A flash fire gutted the rig and living quarters, but the situation was ideal for diver activated well killing techniques. The wild well was killed with a relief well 15 months after the blowout.

CONCLUSIONS

Blowout equipment is much like insurance; the greater the risk to life and property, the greater the cost of offsetting the risk. Despite equipment improvements and increased reliability, catastrophic surface blowouts will occur. Whatever the

Fig. 3–5   *Gas well blowout offshore Australia in 330 ft. of water.*

cause, each situation must be met with an intelligent approach that will lead to the logical control of the well. A prepared plan of action for foreseeable eventualities should be available. This plan, along with the necessary equipment, must be ready before the blowout occurs.

In this section some emergency procedures in the event of offshore blowouts and fires are discussed, and some techniques are described that can be used to bring catastrophic blowouts under control. As operating companies move into deeper waters, the tremendous financial burden and unfavorable public opinion associated with major blowouts should assure that adequate equipment, proper training of operating and contract personnel, and continued testing of equipment and personnel are provided for offshore drilling operations.

REFERENCES

1. Harold H. Davenport, et al, "How Shell Controlled Its Gulf of Mexico Blowouts," *World Oil*, November 1971, pp. 71-77.

2. Carlos Byars, "Burned Platform Comes Back in Gulf Coast," *The Oil and Gas Journal*, July 31, 1972, pp. 96-98.

3. Industrial & Public Relations Department, "Fire on Platform 215-B" Amoco Production Company, 1972.

4. "Chevron Caps Renegade Wells," *Ocean Industry*, May 1970, p. 26.

5. F. J. Schempf, Jr., "Huge Clean-up Force Works on Spill in Gulf," *Offshore*, April 1970, pp. 33-37.

6. *Ocean Oil Weekly*, Vol. 4, No. 25, March 23, 1970.

7. *Ocean Oil Weekly*, Vol. 4, No. 48, August 31, 1970.

8. M. D. Reifel, K. G. Rowley, "Control of Offshore Well Blowouts," Presented at 26th Annual Petroleum Mechanical Engineering Conference, September 1971.

9. F. E. Quon, "Control of Offshore, Onshore, Exploratory Development and Production Wells in the Event of Fire or Blowout," Hurtsteel Products, Ltd. report, October 1971.

10. M. D. Reifel, K. G. Rowley, "Tests Prove Method for Offshore Blowout Control," *The Oil and Gas Journal*, October 18, 1971, pp. 79-88.

10

Annular Blowout Preventers

H. L. Elkins
Hydril Company

Purpose Of Annular Blowout Preventers

The annular blowout preventer (BOP) is the most widely used blowout preventer in the oilwell drilling industry. The purpose of BOP's in general is to close the well bore in the event that the well starts a kick. Ram preventers historically are limited to either blind rams which are closed to control the well (provided no tools are in the well bore) and to a specific pipe size, if pipe is in the hole. This normally precludes closing on tool joints, kellys, etc. If vertical movement of the pipe is required, much effort and duplication of ram sizes is needed to accomplish this operation.

Annular preventers were designed for the express purpose of providing well bore closure under any conditions, regardless of the size or shape of objects in the well bore, or an open hole if the need should arise.

History Of Annular Preventers

Annular preventers in one form or another have been used in the industry for more than 35 years. These can, for purposes of historical discussion, be divided into two types:
1. *Stripper-Type Units*—Preventers whose packoff units were without steel reinforcement.
    a. Hydril Type "CP" (1937). This was the first annular pre-

258

venter to incorporate into one unit the ability to pass tools and drill bits and to close on any shape present in the well bore. It used both externally applied hydraulic pressure and the well bore pressure to effect a closure.
  b. Hydril Type "B" (1938). This was limited to use with externally flush drill pipe and a round kelly.
  c. Regan Type "KFL".
2. *Metal Reinforced Packing Unit Preventers*
  a. Hydril Type "R" (1938). This was the first metal reinforced packing unit used in annular blowout preventers.
  b. Hydril Type "GK" (1947). This is the first of the modern day annular preventers. In 25 years of service, this preventer has become the standard of the industry.
  c. Shaffer "Spherical" (1970). This preventer incorporates the metal reinforced packing unit principle in a short body construction.
  d. Cameron Type "A" (1970). While incorporating a metal reinforced packing unit, the Cameron preventer utilizes a radically new packing unit design and method of operation.
  e. Hydril Type "GL". Incorporates many of the features of the "GK" but is specifically designed for subsea use.

Well Diverters

Well diverters are a special class of large bore preventers with nominal working pressures of less than 1,000 psi. Drilling regulations in many areas today require that adequate blowout protection be provided from the time the well is first spudded until it is completed. In the past, it was standard practice to reduce large conductor pipe to mate with 20 inch blowout preventers. In many instances, the planned casing program required either under-reaming the surface pipe hole or removing the preventers in order to run large-diameter surface casing, or both, thus possibly leaving the hole unprotected. The use of well diverters eliminates the necessity of under-reaming since the well may be drilled to full bore the first time in the hole. This also makes it possible to run large size casing without removing the preventers.

1. *Hydril MSP 29-1/2-500*
   This diverter is capable of full closure from an open bore of 29½ inches. Currently being used on land rigs and platform operations.
2. *Regan 30 inch KFL 1000.*
3. *Regan Riser Diverter—Type "KFD" or "KFDG"*
   This diverter is similar to the "KFL" except that it is adapted for use on marine riser systems. The unit is attached to or built into the rig floor structure on the top of the riser. Misalignment is accommodated by a ball joint below the diverter in the "KFD" version, where the "KFDG" is a gimbal mounted unit. Packoff is achieved down to about 4½ inches on pipe depending on the insert used. The insert must be removed to pass tools.
4. *Vetco Type "HY" Diverter*
   A newcomer to the diverter market, this unit is similar to the Regan "KFD" line in installation, operation, and limitations.

Operational And Design Considerations

1. *Past Usage*
   When one looks back over the history of blowout preventers it becomes apparent that the preventers in use today, both ram and annular, are not too different from those used 25 years ago or more. The principal difference lies in the operating mechanism and not in the concept of the sealing element.

   The basis for most annular preventers on the market today was in part disclosed in a patent application (Knox) filed in 1947. In many respects, all annulars in use today are an offspring of this cluster of ideas. Cameron, in their Type "A," has attempted to make the only real departure from what has been used for many years.

   This is not as bad as it may sound. However, the real need for a different approach has never presented itself. The industry requirements simply have not made such a quest necessary until now.
2. *Present Day Requirements*
   Up until about 1960, the prime consideration in annular

preventer design was the ability of the preventer to establish well bore closure and to maintain this closure against well bore pressure, even in event of the loss of operating fluid pressure. It was this consideration that required the strong self-energizing closing features that characterized all preventers, both annular and ram, up to that time. This feature was not without a penalty; the penalty was reflected in the force required to move pipe through the packing unit with pressure in the well bore. The relatively high contact pressure of the packing unit rubber on the pipe during such movement, in many cases, shortened the life of the packing units.

In the early 1960's, floating drilling operations, as well as a change of operational problems on land rigs, required the movement of pipe through closed packing units as a matter of routine operations. On offshore rigs, the heave of the vessel moved the pipe up and down through the preventers, while for onshore operations the desire to strip pipe in and out of the well under pressure presented the same problems. Subsequently, the self-closing feature has been diminished almost to the zero point to accommodate the change in operational procedures.

The advent of subsea drilling stacks, especially those used by floating vessels, has pointed up some other considerations. One is the overbalance effect of the weight of the mud column in the riser on annular preventers when they are closed, with the net effect being to attempt to open the packing units. Another problem is the extension of the life of the packing unit; this is being approached by the use of new elastomers and better control of the force between the packing unit and the tools in the hole.

Modern day drilling has required the development of new sizes and pressure ratings i.e., 18¾ inch, 5,000 psi; 21¼ inch, 5,000 psi.

Principal High Pressure Annular Blowout
Preventers Available Today

1. *Cameron Type "A"*
    The Type "A" represents a new approach to annular field.

The packing unit is radically compressed by a series of radically acting pistons. This radical layout gives the Type "A" the lowest overall height of the annulars for a given size. To date, this preventer has been built in the 10 inch and 13⅝ inch, 5,000 psi sizes.

2. *Hydril "GK"*

This preventer has been on the market since 1947 and has become the standard in the industry for both onshore and offshore operations. The preventer was designed so that the well pressure will assist the closing action of the packing unit.

3. *Hydril Type"GL"*

Designed for subsea usage, this unit has retained the best features of the "GK" while adding a secondary closing chamber called "balancing chamber" which functions as an independent and fully redundant closing chamber. It can be connected to a continuous pressure supply, or accumulator, to bias the piston to produce fail-safe closure mud-weight compensation in subsea risers, or to decrease the fluid requirements for normal operations of the preventer. The use of this chamber permits close control of packing units to permit safer stripping of tool joints as well as rotation of drill pipes by allowing the closing pressure to be set at optimum packoff tightness for longer packing unit life at any well pressure.

4. *Reagan Type "KFL"*

The KFL utilizes three packer elements to seal off on the drill string: insert packer, main packer, and packer sleeve. Three hydraulic lines are required to operate the KFL, two lines to actuate the insert packer locking dogs, and one line to close or open the packing elements.

The KFL preventer, equipped with surge dampeners to facilitate stripping tool joints for subsea operation, is not a self energizing preventer. When closing the KFL, it is essential that pressure be maintained at 500 psi greater than casing pressure.

5. *Shaffer "Spherical"*

The Spherical preventer is a relatively new entry into the annular preventer market. It utilizes a spherical head covering the top of the packing unit to translate axial loading on the packing unit into radial movement. The overall height of the preventer is somewhat shorter than either the Hydril or Regan but is higher than the Type "A". Shaffer achieved this reduction in height through the use of a short piston and the spherical geometry of the head.

Hydraulic Operation Of The Annular Blowout Preventer

The preventer should be connected to an appropriately sized hydraulic accumulator system that will have 1,500 psi minimum hydraulic pressure available at all times. A manifold with a pressure reducing and regulating valve along with appropriate four-way valving should be used. The normal hydraulic operating pressure range of the Hydril preventer varies from a minimum of 350 psi up to 1,500 psi maximum. Check valves must never be installed in the piping system between the regulating valve and the preventer.

When stripping tool joints through the preventer packing element, the regulating valve must be able to relieve pressure for the packing element to cycle properly. For underwater use of the annular type preventer, it is recommended that 10 gallon accumulators be installed on both the closing and opening chamber hydraulic lines directly at the preventer in order to absorb the pressure surges created by the preventer packing element while stripping drill pipe.

Summary

The annular blowout preventer is one of the most important and useful items of blowout prevention equipment. Proper sizing, operation, care, and maintenance are necessary for optimum use and maximum life. The properly used and maintained annular blowout preventer may be the difference between the satisfactory completion of an oilwell or disaster during blowout conditions.

11

Ram Type
Blowout Preventers

H. L. Elkins
Hydril Company

Currently, there are three manufacturers of ram type blowout preventers: Cameron Iron Works, Hydril, and Rucker-Shaffer.

In 1964 Cameron Iron Works brought onto the market the Type "U" Ram BOP. This BOP was designed specifically for underwater usage. However, it has become the mainstay of the Cameron Iron Works BOP product line. The type "U" BOP is furnished in all sizes and pressure ratings. It is now being used for underwater drilling, including 18¾ inch and 21¼ inch 10,000 psi working pressure. Though the size and pressure rating may vary, the same basic design exists throughout the product line.

Ram change is accomplished by removal of four bonnet bolts on each end of the BOP. Hydraulic pressure is applied to close the BOP. This pressure will cause the bonnets to slide to the open position exposing the rams. Rams are removed by lifting straight up. No unbolding is required. To bring the bonnets back to the closed position, hydraulic pressure is applied to open the BOP. This pressure will cause the bonnets to be pulled back against the BOP body. It is good practice to maintain the opening hydraulic pressure until the bonnet bolts are tightened.

The type "U" ram is oval in shape and is of one-piece construction with a two-piece seal assembly. The face seal contains metal extrusion plates. As the rams close, these extrusion plates force the rubber seal to feed forward and make up for wear or damage. The upper extrusion plate also keeps the seal from flowing up and out while the rams are under pressure. Once the rams are closed, well pressure adds assist to hold the rams closed and will maintain the rams closed should hydraulic pressure be lost.

To change ram seals, the upper seal is pried upward and out of the ram. This will release the face seal which can be forced out of the ram. To replace the seals, the face seal must first be driven completely into position in the face of the ram and then into the ram. The pins in each end of the upper seal lock the face seal firmly into place. Note: The face seal cannot be removed without first removing the upper seal.

The rams are forged alloy steel. The face of the pipe rams are of a hardness greater than the normal drill pipe and tool joints now in use. This allows drill pipe tool joints to be supported by the rams, which is often necessary when drilling from floating vessels. The rams are capable of supporting drill pipe loads of 600,000 lbs. without damage to the rams. However, weights of 150,000 lbs, or greater will cause damage to the tool joint. This damage is generally not enough to cause the tool joint to break. The tool joint should be removed as soon as it is feasible.

The top edge of all pipe rams contains a small counterbore. This counterbore allows for some deformation of the edge of the bore from repeated tool joint suspension without having that edge protrude above the top surface of the ram. Any protrusion above the top of the ram will tend to scratch the sealing surface area in the BOP body each time the rams are operated.

Wedgelocks are used to lock the rams in the closed position. They are operated by hydraulic pressure from the BOP control system and are operated separately, not as part of the open-close hydraulic system for the BOP. On a BOP stack containing several rams, it is standard practice to connect all wedgelocks in parallel hydraulically. When it becomes necessary to close one set of wedgelocks, hydraulic pressure will be applied to all wedgelocks. Due to the design of the wedgelock, it can move into the lock position only when the rams are closed. As long as the BOP rams are in the open position, the ram shaft extension holds the wedgelock in the unlock position not allowing it to move even though hydraulic pressure is applied.

Cameron is currently offering two types of combination blind/shear Rams. The first type was put on the market in 1968 and is in general use on most Cameron subsea BOP stacks. These rams are designed such that the face of the blade in each ram block acts as a sealing surface against the ram packer rubbers.

This design requires that both cutoff pieces of pipe must be removed after the shearing operation to allow the rams to seal the hold as blind rams. The use of these shear rams requires a modification to the BOP in order for hydraulic system pressures of less than 3,000 psi to be used. This modification consists of booster cylinders added to the BOP to double the amount of hydraulic force available to close the BOP. These booster cylinders are installed on the BOP bonnet in the same position as the wedgelocks are normally installed.

Wedgelocks cannot be used with this booster modification. Therefore, a series of pilot and check valves are used to hydraulically lock the rams in the closed position.

Early in 1972 Cameron put a new shear ram on the market. This ram allows the cutting of pipe without the necessity of removing either piece of cut pipe to allow the rams to seal the hole as blind rams. These rams also require less hydraulic system pressure to cut drill pipe and therefore do not require the addition of booster cylinders to the BOP in most sizes. A modification to the BOP is still necessary, however, for the 16¾ inch, 5,000 psi size and smaller sizes. These rams require a longer travel of the ram shaft, therefore, a special bonnet assembly with extended ram shaft travel must be furnished for these rams to be used with the standard type "U" BOP. In this case, the wedgelocks may again be used rather than the hydraulic locking system of check valves.

Another major manufacturer of ram type blowout preventers, Rucker-Shaffer (previously known as Shaffer Tool Works), brought the type LWS Blowout Preventer onto the market in 1961. This BOP was designed primarily for land use, but, with the addition of Automatic Ram Locks, it became very satisfactory for underwater usage. The model LWS BOP underwent major design modifications in 1970 resulting in the model 70 LWS which is designed primarily for underwater usage.

The LWS BOP is furnished in all sizes and pressure ratings for use in underwater drilling including 18¾ inch and 21¼ inch, 10,000 psi working pressure.

The LWS BOP uses a two-piece ram assembly and single piece seal assembly. The two-piece ram allows the use of a floating ram principle whereby the upper seal does not have an interference fit against the upper sealing surface in the BOP body until the ram is almost closed. The rear section of the ram (ram holder) continues to move forward after the main ram blocks have met each other and have stopped, thereby extruding the upper seal tightly against the body of the BOP. Well bore pressure acts against the rams to help effect a tight seal.

The rams are made up of heat treated alloy steel which is stronger than any drill pipe and tool joints being used today. Drill pipe weights of greater than 660,000 lbs. can be supported by tool joints on the pipe rams without damage to the rams. However, drill pipe weights of greater than 400,000 lbs. will cause damage to the tool as it attempts to pull through the rams.

For underwater use, the LWS BOP is furnished with Positive Automatic Ram Locks. The Autolocks are an integral part of the hydraulic operations of the BOP. They are automatic in operation. Each time the rams are closed they are automatically locked. The locks are actuated by the normal open-close system of the BOP.

Ram change on the LWS is accomplished by removing the bonnet bolts and swinging the doors open. The hydraulic system is built into the hinge assembly, therefore, it is not necessary to disconnect any of the hydraulic systems. Hydraulic pressure can be applied to the BOP with the doors open to cycle the ram shafts in and out for inspection. To change the ram seals, the following procedure is used. Once the complete ram assembly is removed from the BOP, remove the two shoulder bolts in the back of the ram holder. This will allow the ram block and seal to be removed from the ram holder. Remove the two seal retaining screws from each side of the ram block. The upper and lower seals can be spread apart and the seal assembly pulled forward off the ram block. Reverse these steps to install the new seal.

It is noted that all components (ram block, ram holder, and seal assembly), are symmetrical, therefore, it is impossible to assemble the pieces upside down, etc. Since the complete ram assembly is symmetrical, it can be placed back in the BOP with either side up. After ram change, the doors are manually swung to the shut position and the bonnet bolts tightened.

In 1962 Shaffer Tool Works developed and patented shear rams for ram type BOP's. Currently, Rucker-Shaffer manufactures two types of shear rams for the LWS Blowout Preventer. The first, or type 70, is very similar to the standard blind rams normally used in the LWS. These rams, however, contain cutting blades. A hardened steel blade forms the center of one ram block and is flushed with the face of that ram block. A tapered blade protrudes from the bottom of the opposite ram block. This blade makes the cut through the pipe and pushes the lower cutoff section, or fish, into a recess beneath the blade in the mating ram block.

It is not necessary to drop the lower fish for the rams to seal. However, it is necessary to remove the upper fish for the rams to completely close with an effective seal. It is not necessary to open the rams to remove the upper fish since it can be pulled out from between the rams with no damage to the seals.

LWS BOP's of 10,000 psi working pressure require no modification for use of the type 70 shear ram. LWS BOP's of 5,000 psi working pressure and lower require installation of larger than standard hydraulic operators in order to have the hydraulic force necessary to shear drill pipe with hydraulic pressures of less than 3,000 psi. Fourteen inch diameter cylinder operators, which are standard on 10,000 psi working pressure LWS BOP's, are readily adapted for use on 5,000 psi working pressure or less BOP's. These operators allow cutting of 5 inch, 19.5 lb. drill pipe within a pressure range average of 2,500 psi. These operators are standardly furnished with Automatic Ram Locks.

In early 1972, Rucker-Shaffer developed a new type of shear ram which is now designated type 72. The type 72 shear ram differs from the type 70 in that it does not require the removal of either the lower or upper fish in order for the rams to seal. Due to design of the ram and blade, much less force is required to shear the pipe. When used with the 14 inch diameter operating cylinder, the type 72 shear ram requires a hydraulic system pressure of only 1,050 psi. When used with the standard 10 inch diameter operator on 5,000 psi working pressure and less BOP's a hyd-

raulic system pressure of less than 2,500 psi is required. Therefore, it is possible to use the type 72 shear ram in the 5,000 psi
working pressure or less BOP's without converting to the larger
operator. Shaffer, however, does recommend use of the larger
operator for greater safety factor.

Both the type 70 and 72 shear rams maintain the floating ram
principle whereby all seals are mechanically set each time the
rams are closed.

The Hydril Company entered the ram blowout preventer market in 1974. First deliveries of ram blowout preventers for subsea
use will be made in 1976. Currently eight sizes and pressure
variations are available or will be by late 1976.

The most distinctive feature of this new BOP is the ram design
itself. The ram is heavier and more durable than any other
available ram. It provides long, reliable service life. The design
has more feedable rubber for sealing than is normally found in
other ram type blowout preventers.

Ram change is accomplished by removing the bonnet bolts
and swinging the bonnets to the open position. Closing hydraulic pressure is applied extending the ram from the bonnet
for easy removal. Once the ram is removed from the BOP the
front packer and upper seal assembly may be removed by first
rotating the ram assembly screws to the left until they tsop, then
lift the ram upper plate out of the ram body.

Pipe rams allow hang off of up to 600,000 lbs. of drill pipe
weight on a taper tool joint while maintaining a seal to rated
working pressure. Shear rams are available to shear all common
size drill pipe and seal on the open hole without removal of
either upper or lower piece of cut off pipe.

For subsea use the Hydril ram BOP has a unique
multiposition/automatic ram lock which gives optimum ram
front packer feed as rams and/or packers wear the usage.

12

Choke/Kill Valves

H. L. Elkins
Hydril Company

All underwater BOP stacks are equipped with valve manifolds to control access to the choke and kill lines attached to the marine riser. All BOP stacks, with the exception of some low pressure 20 inch single annular BOP stacks, are equipped for both choke and kill line operation.

A basic bore size and pressure ratings of 3⅛ inch, 5,000 psi working pressure and 3-1/16 inch, 10,000 psi working pressure have been standardized by the various manufacturers. There has been discussion, however, of going to 4-1/16 inch, 10,000 psi working pressure components for use with the larger BOP stacks, i.e., 18¾ inch and 21¼ inch, 10,000 psi working pressure sizes.

Valves used are hydraulically operated, fail-safe closed. Hydraulic pressure is required to open the valve only. Closing the valve is accomplished by spring force within the valve operator and by some means of assist from line pressure. An inherent problem of all hydraulic operated fail-safe valves is the reaction force caused by hydrostatic head always present on the opening pressure side of the hydraulic piston in the valve operator.

This pressure would tend to hold the valve open by balancing the spring force on the underside of the piston which is the closing force for the valve. It is necessary then to allow the hydrostatic head to also react on the pressure close side of the hydraulic operator piston. This is accomplished for the various makes of valves by either oil filled seal pots or oil filled bellows chambers. This allows the valves to operate at various depths under water without adverse effects from the hydrostatic head. Figure 3-6 shows a cutaway of a typical hydraulic fail-safe close gate valve.

The major manufacturers of choke/kill valves are: Cameron Iron Works, Hydril Company, Oil Center Tools Division of FMC Corporation, Rockwell-McEvoy, Rucker-Shaffer, and the W-K-M Division of ACF Industries. With the exception of Hydril Company, who uses a ball valve design, the other manufacturers all furnish various gate valve designs. All manufacturers now offer an angle or cross type valve body to minimize the number of connections required to install the choke/kill line manifolds on the BOP stacks.

Much discussion centers around the proper placement of the choke/kill line outlets on the BOP stack. Figure 3-7 shows a typical setup for a three ram BOP stack. Two valves are used for each choke/kill line for redundant control. The choke line is located below the bottom rams. This will allow circulation beneath both sets of pipe rams above. The kill line is located between the two lower sets of rams. This allows the fill line to be used for circulation as a choke line if the upper pipe rams are closed. Should it be necessary to cut the drill string with the shear rams, the drill pipe can be hung off and sealed on the bottom rams. Circulation is possible by pumping down the kill line and flowing out the choke line.

Figure 3-8 shows the typical setup for a four ram BOP stack. The choke line is located between the two lower sets of rams. The kill line is located between the two top sets of rams. This allows circulation in the same manner as a three ram BOP stack. No circulation is possible beneath the bottom rams. This may seem to be a waste of the bottom rams. However, this is instead a very good safety feature. Should the choke line become washed out or otherwise damaged, the bottom rams may be closed, thereby shutting off pressure to all components above the bottom rams.

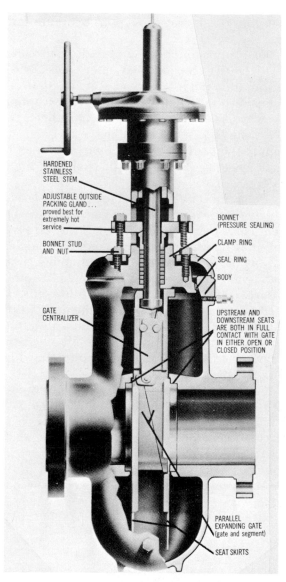

Fig. 3–6   *Cutaway of typical hydraulic gate valve.*

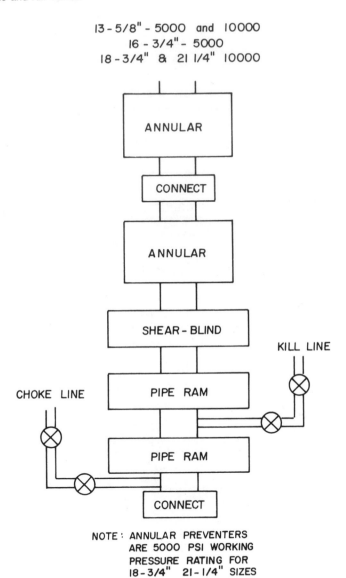

Fig. 3–7 *Subsea blowout preventer stack.*

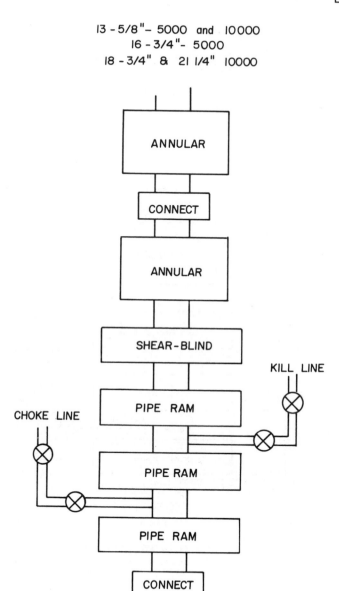

13 - 5/8"- 5000 and 10000
16 - 3/4"- 5000
18 - 3/4" & 21 1/4" 10000

Fig. 3–8 *Subsea blowout preventer stack.*

13

Unitized Blowout Preventer Stacks

H. L. Elkins

Hydril Company

Depending on the casing program and expected drilling conditions, marine drilling operations can be carried out by using two separate riser sizes and corresponding blowout preventer stacks, or by using one single riser size and blowout preventer stack.

When using two risers and two blowout preventer stacks, the system consists normally of a 24 inch riser and 20 inch blowout preventer stack. This blowout preventer stack is installed on the 20 inch surface casing. After a 17½ inch hole has been drilled and 13⅜ inch casing run to the desired depth and cemented, the 20 inch blowout preventer stack is removed and a 13⅝ inch blowout preventer stack and 16 inch riser is installed on top of the 13⅜ inch casing.

The single riser and blowout preventer stack system originally consisted of an 18⅝ inch riser and a 16¾ inch blowout preventer stack. This blowout preventer stack is attached to the 16¾ inch wellhead which is installed on top of the 20 inch surface casing. A 15½ inch hole is drilled through the stack, and then is underreamed to 17½ inches to accommodate the 13⅜ inch casing string which is then run to the desired depth and cemented.

An alternative to the single 16¾ inch blowout preventer stack is a 21¼ inch blowout preventer stack with a 24 inch riser, or an

18¾ inch stack with a 20 inch riser. With this sytem, no under-
reaming is required after the surface casing has been set. All
drilling and completion operations can be performed through a
single blowout preventer stack.

Rated Working Pressure

The rated working pressure of the blowout preventer stack
must be adequate to confine the maximum anticipated pressure
to which it may be exposed. When the two blowout preventer
stack system is used, the working pressure rating of the 20 inch
blowout preventer stack is usually 2,000 psi. Ram type preventers of 3,000 psi working pressure rating may be used, if desired,
although annular preventers of this size are presently not rated
higher than 2,000 psi.

The 13⅝ inch BOP stack which follows the 20 inch BOP stack
has a working pressure rating of either 5,000 psi or 10,000 psi
depending on the expected maximum pressures to which the
equipment may be exposed.

In single stack systems the 16¾ inch BOP stack is normally
rated at 5,000 psi working pressure. At present, ram type preventers in this size are available at higher working pressure ratings
of 10,000 psi. The 18¾ inch and the 21¼ inch blowout preventer
stacks are now available with 10,000 psi working pressures.

Stack Arrangements

Figures 3-9 and 3-10 illustrate 20 inch, 2,000 psi BOP stacks.
Figure 3-10 shows a similar arrangement where the annular
preventer is combined with two ram type preventers.

Figure 3-11 shows a BOP stack for various sizes and working
pressure ratings of 5,000 psi and higher. This stack is a combination of two annular preventers with three ram type preventers.
Connectors are located on the bottom of the ram preventers and
between the annular preventer to be retrieved on the riser for
repair of change of element leaving the rest of the BOP stack
undisturbed. In operation, the upper annular preventer (sometimes called "working preventer") is the first to be used.

Figure 3-11A is a cheaper alternative to the arrangement of
Figure 3-11. The three ram type preventers are combined with

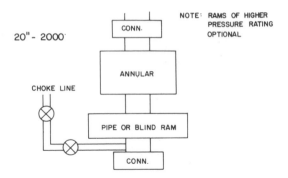

Fig. 3–9 *Subsea blowout preventer stack.*

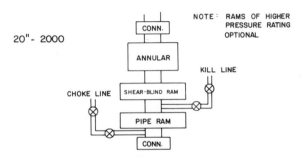

Fig. 3–10 *Subsea blowout preventer stack.*

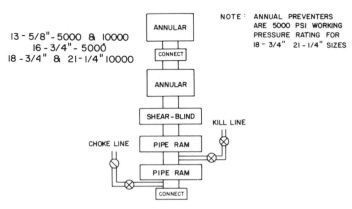

Fig. 3–11 *Subsea blowout preventer stack.*

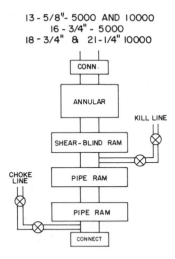

Fig. 3–11A   *Subsea blowout preventer stack.*

only one annular preventer. The redundancy of this system is obviously inferior to the arrangement in Figure 3-11. Figure 3-12 has an identical annular preventer arrangement to Figure 3-11. Here, however, the annular preventers are combined with four ram type preventers for additional redundancy and more flexibility. This arrangement is now frequently used in deep water drilling operations.

Figures 3-12A and 3-12B are alternative arrangements to the one shown in Figure 3-12. The arrangement in Figure 3-12A is less redundant because of the omission of the second annular preventer. The arrangement in Figure 3-12A is less flexible because the upper annular preventer cannot be retrieved separately on the riser.

Location of Rams and Choke/Kill Lines

Rams (including shear/blind rams) and choke and kill line connections are readily interchangeable in a BOP stack. Before the stack is submerged, their relative position in the stack can easily be altered to suit changed conditions or the operator's

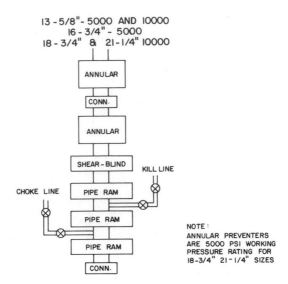

Fig. 3–12   *Subsea blowout preventer stack.*

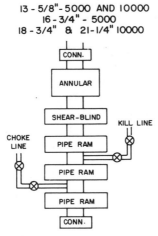

Fig. 3–12A   *Subsea blowout preventer stack.*

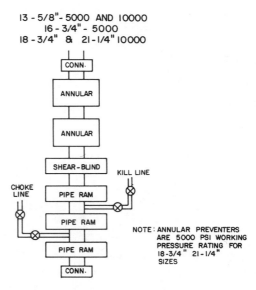

Fig. 3–12B   *Subsea blowout preventer stack.*

preference. As an example, Figure 3-13 shows one of the most common arrangements of rams and choke and kill lines for a subsea BOP stack. The advantages of this arrangement are:
  a. Shear/blind rams in the top preventer leave at least two (and if the spacing between the top preventer and second from the top is sufficient, three) sets of pipe rams to hang the drill pipe on.
  b. Upper circulating line between top rams and second from top allows circulation through lower cut portion of drill pipe regardless of which ram the pipe is hung on.
  c. Lower circulating line between bottom rams and second from bottom leaves no outside connection below bottom rams. When the bottom rams are closed and the well is abandoned, this arrangement eliminates the possibility of failure due to damage to vulnerable protruding connections.
The disadvantages of this arrangement are:
  a. When rams are closed on pipe, circulation out is possible only through the lower circulating line.

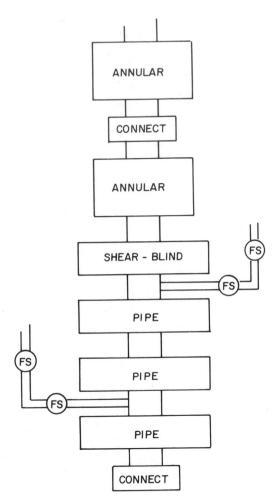

Fig. 3–13   *Subsea blowout preventer stack.*

b. When bottom rams are closed, no circulation is possible.

It is possible that other ram and circulating line arrangements are possible and their relative advantages will be the determining factor when selecting the rams and circulating line arrangement for a specific condition.

14

Blowout Preventer
Operating Procedures

H. L. Elkins
Hydril Company

The following covers the use of the blowout preventer (BOP) stacks and equipment used when drilling from a floating vessel. Figure 3-14 shows typical BOP stack with the major components normally used in a typical BOP stack. Though the size and ·pressure ratings may vary, all major BOP stacks in use for subsea drilling today contain the major components shown. These major components are: (1) hydraulic wellhead connectors; (2) ram type BOP's; (3) annular type BOP's; (4) hydraulic or electrohydraulic control system for actuation of all the components on the BOP stack; (5) a four-post guide frame used to guide the BOP stack down the guidelines to the wellhead landing base and line up the BOP stack with the wellhead for proper seating.

Ram BOP's contain pipe rams sized to fit the drill pipe being used. In some cases, rams are installed to fit the various size casing strings being landed through the BOP stack. At least one chamber in the ram BOP's will contain blind rams or, as in common practice now, a combination blind and shear ram assembly. The combination blind/shear rams operate under standard conditions as blind rams to close off the well bore at any time that drill pipe or tools are removed from the hole giving a complete closure of the hole.

However, in an emergency these rams can be closed effectively on the drill pipe severing the drill pipe in an emergency

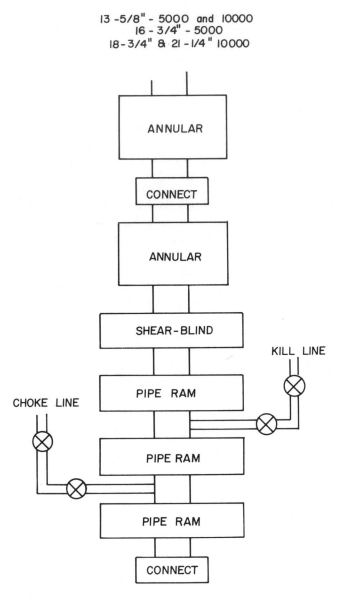

Fig. 3–14 *Subsea blowout preventer stack.*

situation when it is not feasible to remove pipe from the hole. For instance, if anchor chains or lines break while drilling and the marine riser must be immediately removed to keep from breaking off from the BOP stack, or if the possibility of a blowout going uncontrolled through the drill pipe exists, the drill pipe must be severed to stop the blowout at the BOP stack rather than at the surface.

The annular BOP, however, is designed to close off any tools that may be in the hole. In most cases, it is drill pipe, but may include casing, tool joints on drill pipe, drill collars, etc. In an emergency the element of the annular BOP can be completely closed on the open hole in the same manner as the blind rams in the ram preventers are concerned. In subsea drilling, the annular BOP is really the heart of the BOP stack. This preventer is the most versatile and most used preventer in subsea drilling.

Unlike drilling on land where visibility allows the location of the kelly or the drill pipe and tool joints to be determined, drilling through a subsea BOP stack does not allow the location of the tool joints to be determined in relation to the BOP stack. It is not feasible to close the pipe rams and ram preventer without knowing the location of the tool joints. If an attempt is made to close the pipe rams and the rams close on a tool joint, the BOP could possibly be damaged. Damage could involve the rams and/or the tool joint to the extent that it may part, dropping the drill string to an emergency situation.

This leaves only one alternative. The annular preventer should be closed first in all cases when necessary to close the BOP on pipe. Since the annular BOP has the ability to close on both tool joints and drill pipe, there is no danger of damaging the BOP or the tool joint itself. Therefore, the annular preventer is used both to close off the hold and to locate tool joints. The forces required to strip a tool joint through the annular BOP are much greater than the forces required to strip drill pipe up or down through the element of the annular BOP.

The drill pipe can then be raised and, with a noticeable change in the weight of the drill string, will indicate a tool joint passing through the element of the annular BOP. As the tool joint is pulled up through the element of the annular BOP, it is positively located. Then it is possible to safely close the ram BOP on drill pipe, not on the tool joint.

Since a floating vessel is being used, the drill pipe will be in continual action at all times with the heave or movement up and down of the drilling vessel caused by the sea. When the BOP is closed, the pipe must still be free to move up and down in order to follow the heave of the vessel. In the past, the stripping life of the BOP has been very minimal. Due to its inherent design features, the annular BOP has been able to strip a greater footage of pipe than the ram preventers.

When feasible, a set of pipe rams should be closed and the drill string lowered until a tool joint is supported on top of the pipe rams. This is possible only when the drill pipe is being handled in elevators on the rig—not when the kelly is attached.

15

Subsea Blowout Preventer Control Systems

John D. McLain and
Darrell G. Foreman

C. Jim Stewart & Stevenson, Inc.

Next to the blowout preventers, the most important component for well control in floating drilling is the system that
monitors and controls the behavior of the subsea BOP's from the
drilling rig.

From 1955 to 1963, the control system design used in subsea
drilling was basically a land rig or "closed-type" system. A
hydraulic power unit provided fluid to a shipboard mounted
control valve manifold. Hydraulic power lines were run from
this manifold directly to each BOP stack function. These hydraulic lines were either run down the riser, were an integral
part of the riser, or were huge independent hose bundles. Actuation of a control valve directed fluid to the respective stack
function. The opposite function on that ram or other stack component discharged back through the respective power line,
through the shipboard mounted control valve, and into the
reservoir of the hydraulic power unit.

The drilling water depths during the initial years of this
period were relatively shallow and the use of the field proven
land rig components was a natural design decision. Closing
times on the blowout preventers were reduced by using larger

I.D. power hoses from vessel to subsea stack. Koomey Division of Stewart and Stevenson designed the first 3,000 psi working pressure accumulator system during this period. Larger volumes of fluid could be stored under pressure in less space, and with the addition of a pressure reducing and regulating valve, actual working pressure of the control fluid could be maintained within the limits of the stack components.

As the water depths increased, serious problems developed with this type system. One, the huge hose bundles used between vessels and BOP stack were not adequate due to vulnerability to currents and the forces of the sea. Also shipboard handling procedures were long and costly. However, to reduce overall O.D. was impossible because with the greater water depths came dangerously longer closing times for critical BOP functions.

In March, 1963, a new concept was designed and put into operation by Payne Manufacturing Company. Subsea pilot operated control valves were installed on the BOP stack. Power fluid was directed to this manifold of subsea valves from the vessel through a single 1 inch hose. One-eighth inch I.D. hoses were used to transmit hydraulic pilot signals from the vessel to the subsea control valves. Additionally, the subsea control valves were designed to "vent to sea" the discharging fluid from a ram or stack component when actuated. Closing times were drastically reduced. The subsea control valves were installed in a retrievable connector called a pod.

Thus, the advent of what is called the "open type" system. This is the most commonly used type system today. Koomey, Cameron-Payne, Valcon, and Hydril Company offer this pod type system. Hydril Company offers the pod in electro-hydraulic form only. Electro-hydraulics will be covered later in detail.

In the past five years much has been done to reduce the actuation or closing times on subsea BOP stacks. By 1968-1969 drilling water depths had gradually increased to a 600 foot maximum. Then, a major oil company outfitted two floating drilling rigs in the Santa Barbara Channel with subsea equipment capable of drilling in up to 1500 feet of water.

Closing times, even with the subsea pilot operated pod valve systems as available, were no longer acceptable simply because of the greater water depths. Kick detection on a floating drilling rig is far more difficult than on a land rig. The flow indicator commonly used on land to monitor flow out of the bell nipple is not accurate on a floater, because the rig's heave causes the flow out of the bell nipple to vary erratically. Also, an increase in the rate of penetration is difficult to detect because the bumper subs must "drill off" before a substantial change in rate of penetration is detected.

One of the other commonly used methods of kick detection, the pit level indicator, is also affected by the vessel motion. Because of the kick detection problems on a floating drilling vessel more extraneous fluids can enter the well bore and rise farther up the well bore before being detected than on a land rig. To offset the lack of speed and accuracy in kick detection, coupled with the greater water depths, a faster actuation of the subsea BOP's was essential.

Koomey introduced the first solution to reduction in BOP closing times in 1969 in the Santa Barbara Channel. Previous pilot hoses were all ⅛ inch I.D. Koomey's solution was a hose bundle of mixed pilot hose sizes—⅛ inch continued to be used on functions not requiring speedy reaction times (surface command to subsea pod valve shifting time), and a new ³/₁₆ inch I.D. pilot hose on stack functions requiring fastest possible reaction times (BOP's). Construction details directly responsible for this faster reaction time with ³/₁₆ inch I.D. hoses and discussed in the following BOP control systems component breakdown.

Today, acceptable closing times can be achieved by proper design and installation of the control system and the related piping for the hydraulic power fluid. However, fast closing times alone are not the only factor to be considered in designing a control system on a floating rig. Overall safety in component layout must be given prime consideration.

Today, there are three basic component layout designs offered. Figure 3-15 shows the most basic and least expensive. However, it is also the least acceptable by the industry today. In this system, the Driller's Panel is the shipboard mounted hydraulic pilot valve control manifold. Located at the driller's position on the rig floor, it is in the most hazardous area of operation on a drilling rig. Destruction or damage to this panel destroys

Koomey Hydraulic Control System With *Hydraulic* Operated Driller's Panel

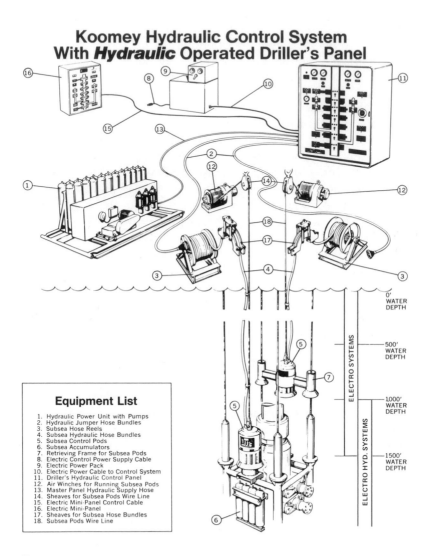

Equipment List

1. Hydraulic Power Unit with Pumps
2. Hydraulic Jumper Hose Bundles
3. Subsea Hose Reels
4. Subsea Hydraulic Hose Bundles
5. Subsea Control Pods
6. Subsea Accumulators
7. Retrieving Frame for Subsea Pods
8. Electric Control Power Supply Cable
9. Electric Power Pack
10. Electric Power Cable to Control System
11. Driller's Hydraulic Control Panel
12. Air Winches for Running Subsea Pods
13. Master Panel Hydraulic Supply Hose
14. Sheaves for Subsea Pods Wire Line
15. Electric Mini-Panel Control Cable
16. Electric Mini-Panel
17. Sheaves for Subsea Hose Bundles
18. Subsea Pods Wire Line

0' WATER DEPTH

500' WATER DEPTH

1000' WATER DEPTH

1500' WATER DEPTH

ELECTRO SYSTEMS

ELECTRO HYD. SYSTEMS

Fig. 3–15 *Koomey hydraulic control system with hydraulic operated driller's panel.*

the entire BOP control system's capabilities. An electric remote panel, usually located in the tool pusher's office, gives a second point of control under emergency conditions as long as the Hydraulic Driller's Panel is operable.

Figure 3-16 shows the shipboard mounted hydraulic pilot valve control manifold located in a safe yet accessible area and remotely operated by an Air Master Panel located at the driller's position. Remote operation is accomplished by sending an air signal through air tubing to an air operator mounted on the hydraulic pilot valve mounted on the pilot valve manifold. Operation of this valve sends the hydraulic pilot signal through the hose bundle to the subsea control valve located in the pod on the BOP stack. An electric remote gives a third point of control. Loss of the Air Master Panel does not negate the BOP control system and there remain two points of control.

Figure 3-17 shows the BOP control system most preferred by the industry today. The pilot valve control manifold is located in a safe area and remotely controlled from the driller's position by an Electric Driller's Panel. This panel is totally explosion-proof and has two distinct advantages over the previously described Air Master Panel: (1) it does not require the multi-tube air cables run between driller's position and pilot valve control manifold, (2) most important, it gives immediate response from command at Electric Driller's Panel to pilot valve shift at Hydraulic Manifold.

On most Air Master Panel installations a time lag of from two to five seconds is required from command to pilot valve shift at Hydraulic Manifold depending on distance and length required on interconnecting air cables.

Remote operation from the Electric Driller's Panel is accomplished by sending an electric signal from the Electric Driller's Panel to an air solenoid valve located in an explosion-proof box on the Hydraulic Control Manifold. The air from the solenoid actuates the air operator on the hydraulic pilot valve sending a hydraulic pilot signal through the hose bundle to the subsea control valve located in the pod on the BOP stack. Again, an electric remote gives a third point of control. An emergency power pack is provided with the system that automatically gives battery reserve power to the two remote panels should rig power

Koomey Hydraulic Control System
With *Air Operated* Driller's Panel

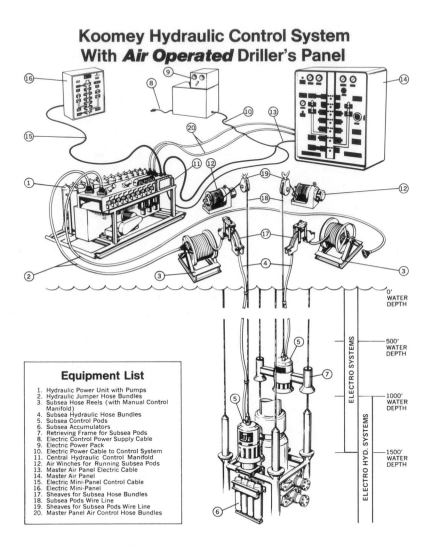

Equipment List

1. Hydraulic Power Unit with Pumps
2. Hydraulic Jumper Hose Bundles
3. Subsea Hose Reels (with Manual Control Manifold)
4. Subsea Hydraulic Hose Bundles
5. Subsea Control Pods
6. Subsea Accumulators
7. Retrieving Frame for Subsea Pods
8. Electric Control Power Supply Cable
9. Electric Power Pack
10. Electric Power Cable to Control System
11. Central Hydraulic Control Manifold
12. Air Winches for Running Subsea Pods
13. Master Air Panel Electric Cable
14. Master Air Panel
15. Electric Mini-Panel Control Cable
16. Electric Mini-Panel
17. Sheaves for Subsea Hose Bundles
18. Subsea Pods Wire Line
19. Sheaves for Subsea Pods Wire Line
20. Master Panel Air Control Hose Bundles

0' WATER DEPTH

500' WATER DEPTH

1000' WATER DEPTH

1500' WATER DEPTH

ELECTRO SYSTEMS

ELECTRO HYD. SYSTEMS

Fig. 3–16　*Koomey hydraulic control system with air operated driller's panel.*

Koomey Hydraulic Control System
With *Electric* Operated Driller's Panel

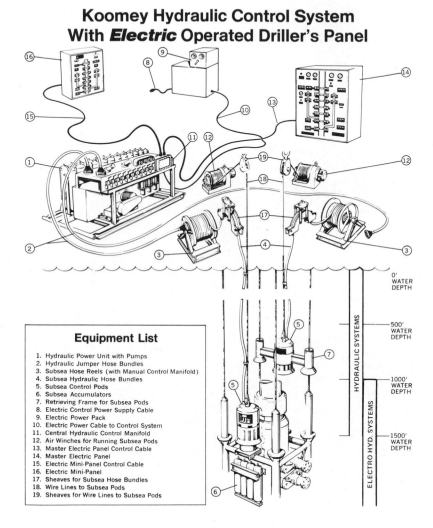

Equipment List

1. Hydraulic Power Unit with Pumps
2. Hydraulic Jumper Hose Bundles
3. Subsea Hose Reels (with Manual Control Manifold)
4. Subsea Hydraulic Hose Bundles
5. Subsea Control Pods
6. Subsea Accumulators
7. Retrieving Frame for Subsea Pods
8. Electric Control Power Supply Cable
9. Electric Power Pack
10. Electric Power Cable to Control System
11. Central Hydraulic Control Manifold
12. Air Winches for Running Subsea Pods
13. Master Electric Panel Control Cable
14. Master Electric Panel
15. Electric Mini-Panel Control Cable
16. Electric Mini-Panel
17. Sheaves for Subsea Hose Bundles
18. Wire Lines to Subsea Pods
19. Sheaves for Wire Lines to Subsea Pods

0' WATER DEPTH

500' WATER DEPTH

1000' WATER DEPTH

1500' WATER DEPTH

HYDRAULIC SYSTEMS

ELECTRO HYD. SYSTEMS

Fig. 3–17   *Koomey hydraulic control system with electric operated driller's panel.*

be lost. The following is a breakdown of each component in a hydraulic BOP control system with recommended sizing specifications and formulas.

Pump-Accumulator Unit

This unit consists of the hydraulic reservoirs, mixing/proportioning system for the power fluid, pumps and accumulators. Sizing of all the individual components is directly related to the size and working pressure of the BOP stack to be controlled.

Blowout preventer control systems, like other hydraulic systems, need a high quality fluid to perform their operations satisfactorily. The hydraulic fluid lubricates all the valves in the control system and acts as a corrosion inhibitor for subsea internal parts. Most drilling contractors have been using water soluble cutting oils mixed at various concentrations for their BOP control fluids. Recent domestic legislation regarding the dumping of any oil base fluid in our offshore waters, coupled with the problems encountered with existing fluids, have prompted Koomey, in conjunction with a major chemical company, to develop a nonpullutant hydraulic fluid for BOP control systems. Cameron-Payne now has a similar fluid on the market.

One of the main problems encountered with the old fluids is the formation of undesirable precipitates when the fluids become contaminated, most notably with sea water. A precipitation problem is also created when hard make-up water is used. The precipitates formed decrease the overall efficiency of the system and have, on occasion, completely plugged the pilot lines in the hose bundles. All of the hydraulic fluids must be compatible with ethylene glycol to prevent freezing.

Regardless of the fluid used, a 40 micron filter system should be used downstream of the pumps. Filters should be connected in parallel so one can be cleaned without closing down the entire system. The make-up water used on a rig should come from a distillation unit on board. Water brought from shore on a workboat should not be used. Periodic flushing of all hoses is also good practice.

The hydraulic fluid previously mentioned is pumped from the mixing reservoir to accumulators, which are nothing more than sources of stored hydraulic energy. There are two basic types of accumulators, the bladder type and the guided float type. Both types used today are precharged with nitrogen to a given pressure (1,000 psi in 3,000 psi working pressure hydraulic control systems) then filled with hydraulic fluid to some higher pressure (3,000 psi), compressing the gas. The gas acts as a spring and forces the hydraulic fluid out of the accumulator.

In the guided float type accumulator, the gas actually comes in contact with the hydraulic fluid. Both Cameron-Payne and Koomey offer the guided float type accumulator in both spherical and cylindrical forms. Koomey recommends and offers a bladder type accumulator for positive separation of gas precharge and hydraulic fluid. The bladder type is also field repairable whereas the guided float type is not.

All accumulators behave in accordance with the universal gas law:

$$\frac{PV}{T} = \frac{mR}{M}$$

$P =$ Pressure
$V =$ Volume
$T =$ Absolute temperature
$R =$ Universal gas constant
$m =$ Mass
$M =$ Molecular weight

(1)

For a given mass, the right-hand side of Eq. (1) will have a constant value regardless of the PVT interaction. If we assume that the accumulator operates isothermally (temperature of the gas remains constant during compression or decompression), the P-V relationship can be written as:

$$P_1 V_1 = P_2 V_2 \tag{2}$$

where the subscripts designate two different sets of conditions.

Some accumulators during charging and discharging tend to act adiabatically. In this condition, the gas does not lose or gain any heat during compression or decompression, and consequently temperature varies. If heat cannot enter or exit, Eq. (2) will not describe the reaction. Separator types of accumulators, especially bladder and diaphragm types, exhibit this type of action. The following equation must be used to describe the pressure volume discharge characteristics:

$$P_1 V_1 \, \gamma = P_2 V_2 \, \gamma \qquad (3)$$

where γ is the ratio of the specific heat at a constant presure to the specific heat at a constant volume for a given gas. Nitrogen is used as the precharge gas because it is an inert gas. For nitrogen, γ is 1.40. When any accumulator discharges, it will react in accord with Eq. (2), (3), or a combination of both.

All accumulators should be precharged with nitrogen, which provides a safe source of potential energy. In an accumulator, the higher the precharge, the less fluid volume the accumulator can hold.

Sizing of the BOP control system components is directly related to the size and working pressure of the BOP stack to be controlled. Once the stack has been defined the first component to be sized should be the accumulator capacity. Koomey recommends, as do most major operators, the accumulator capacity to provide fluid to open and close all ram type preventers and the primary annular preventer plus 50% reserve.

A quick method to determine the volume of accumulators on a given BOP stack is to determine the fluid volume required to open and close the preventers w/50% reserve. Multiply this figure by 1.98 to determine the total accumulator volume.

A portion of the accumulator capacity should be placed on the BOP stack. The reason for this is to reduce closing times. By placing a portion of these accumulators as close to the pod valves as possible the friction loss experienced with all accumulators on surface is reduced. In testing a 16¾ inch 5,000 psi working pressure BOP stack during rig-up, the subsea accumulators were isolated from the system and the annular preventer closed and timed. Hose bundle length was 750 feet with approximately 100 feet of jumper hose bundle between hydraulic manifold and hose reel. The subsea accumulators were then opened to the system and the test repeated. With no subsea accumulators the closing time on the annular preventer was 30 seconds. With subsea accumulators the closing time was reduced to 14.8 seconds.

Further testing is being done at present to determine the accumulator volume required subsea to give optimum closing times on any given size preventer at any given water depth.

When the subsea accumulators are precharged, the hydrostatic pressure due to the working water depth must be added to the precharge pressure to compensate for the loss of fluid caused by the hydrostatic head. There is a substantial difference in the volume change as water depth increases.

The precharge on all accumulators should be checked every time the stack is on the surface. Those accumulators permanently mounted on board the vessel should have their precharge checked at least once a week. An easy method for checking the surface accumulators is to isolate the individual banks of accumulators and bleed them down separately. As you are bleeding the pressure off, observe the pressure gauge on the unit labelled "Accumulator Pressure." This gauge will slowly fall to the precharge pressure and then fall immediately to zero as the accumulators close. If the gauge slowly falls below 1,000 psi before dropping rapidly to zero, one or more accumulators is not fully precharged to 1,000 psi. The subsea accumulators must be checked individually with a pressure gauge each time the stack is on the surface.

Pumps should be provided with a dual power source for safety. The accepted standard is to provide both air powered and electrically driven pumps on the accumulator unit. Both the air powered and electrically driven pumps should be 3,000 psi working pressure. In sizing the pumps, a recovery time from 0 to 3,000 psi of the accumulator volume should be 10 to 15 minutes. Two middle-sized pumps generally are preferred to one master pump for the safety in redundancy. The electric pumps should always be connected to an auxiliary generator for use in emergency conditions.

The mixed fluid reservoir should be sized to hold adequate fluid to completely charge the accumulators from 0 to 3,000 psi. The mixing system should be designed to have a mixing rate at least equal to the maximum pumping rate of the pumps.

Hydraulic Control Manifold

This is truly the "heart of the system." As previously discussed, it is a recommended and accepted practice now to place this item in a safe, yet accessible, location on the vessel. Contained in the hydraulic manifold are pilot control valves (normally one per hydraulically actuated stack component) complete with air operators and hydraulic pressure gauges for the various pressures required on the BOP stack, complete with pressure transmitters to convert these hydraulic pressures to either air signals or electric signals and transmit them to either the Air Master Panel or the Electric Driller's Panel.

Also located on this item are the pilot regulators that remotely control the subsea hydraulic regulators located in the pod. These pilot regulators are remotely controlled from the Driller's Panel (Air or Electric). There are normally three regulators on each Hydraulic Control Manifold—one for the ball joint, one for the annular preventer(s), and one for the remaining stack functions. Explosion-proof boxes containing pressure switches are included for remote panel indicating light operation, and air solenoid valves enable remote operation of the pilot control valves and pilot regulators.

Hydraulic Hose Bundles

Hydraulic hose bundles are used to transmit pilot signals and power fluid from the Hydraulic Control Manifold to subsea pods. At present there is only one accepted manufacturer of hydraulic hose bundles for use in subsea BOP control systems: Samuel Moore Company, maker of synflex hose. Synflex hose consists of an inner nylon tube wrapped in from one to four braided sheets, then covered with polyurethane. The two sizes of pilot hoses used in the subsea bundles are the ⅛ inch I.D. 3130 series and ³/₁₆ inch I.D. 3300 series.

The ⅛ inch pilot hose inner nylon tube is wrapped in two braided sheets, while the ³/₁₆ inch pilot hose inner nylon tube is wrapped in four braided sheets. This allows the ⅛ inch pilot hose to have a greater expansion characteristics than the ³/₁₆ inch pilot hose. The greater the expansion, the more fluid is required to generate a signal and completely actuate the subsea pod valve.

In May 1974, Samuel Moore introduced a new $3/16$ inch hose. It has a maximum O.D. of 0.358 inches where the old $3/16$ inch hose was 0.523 inches O.D. The O.D. of the new $3/16$ inch hose compares very closely to the old $1/8$ inch hose which is 0.334 inches. The new $3/16$ inch hose allows the hose bundle to be all $3/16$ inch hoses and have a smaller diameter than a mixed hose bundle formerly did. The smaller O.D. allows more hose on the same size hose reel and improves the overall signal time for all functions on a blowout preventer stack.

Subsea Pods

The subsea pod contains the pilot operated control valves and pilot operated regulators required to direct the hydraulic fluids to the various stack functions. Pods may be either of the retrievable or nonretrievable type. The advantages of the retrievable type pod greatly outweigh the nonretrievable type, and it is the most commonly accepted by the industry. The retrievable male portion of the pod contains all pod valves, regulators, and the hose bundle junction box. Should a pod valve, regulator, or hose bundle malfunction, it is less costly to retrieve the pod than to retrieve the riser and upper stack assembly.

Retrievable pods are of two different designs. Koomey offers the male portion in the form of a tapered single stab with packer type face seals. Cameron-Payne offers the male portion in the form of multiple pin type stabs looking down from the valve compartment. Seals on these pin stabs are optional: O-ring or chevron packing. Koomey uses a patented subplate mounted (SPM) poppet type pod valve. The valve block is drilled, tapped and ported to contain these valves. Cameron-Payne offers a shear seal type pod valve. These valves are manifolded in a compartment and piped direct to the pin type stabs. Both Koomey and Cameron-Payne manufacture subsea regulators. A readback is strongly recommended downtream from the subsea regulator to insure that the proper pressure is attained.

All subsea control systems being manufactured today have 100% redundant hose bundles and pods for safety.

Shuttle valves are used to isolate the pod not in use. Redundancy ends at the shuttle valve. It is strongly recommended that the shuttle valves be piped directly into the part on the stack function rather than packing them at one location and then running hose to the function.

The stack plumbing should be sized and run carefully to eliminate all flow restrictions.

Electro-hydraulic Control System

The electro-hydraulic control system is similar to the hydraulic system, except that an electric signal is sent subsea to a solenoid valve which supplies hydraulic pilot pressure to the subsea control valves. One of the main advantages in the electrohydraulic system is the reduction in signal time to almost zero for any water depth. The electro-hydraulic control system costs are higher than the costs of comparable all-hydraulic systems for shallow water application, but the reverse begins to be true for water depths between 1500 and 2000 feet. Another advantage of the electro-hydraulic system is that it has more readback capabilities.

There are now two types of electro-hydraulic control systems offered by all control system manufacturers. One is the multi-wire electro-hydraulic and the other is the multiplex electro-hydraulic. Multi-wire means there is at least one, or perhaps two, wires per solenoid. The multiplex control system uses from six to twelve wires for all the solenoids. This means the signal for each solenoid must be a coded signal for that particular solenoid. The multi-wire electro-hydraulic control system uses more wires but has fewer electronic components, whereas the multiplex control system uses more electronic components on the surface and subsea, but has fewer wires in the cable.

In 1970 when the current electro-hydraulic control systems came out on the floating drilling rigs, the systems were all multi-wire control systems. The reason for going to the multi-wire system was that the industry did not feel it was ready for sophisticated electronics on a drilling rig. All electro-hydraulic systems worked after modifications to the original electro-hydraulic cable. The biggest problem with the electro-hydraulic was the cable; the remainder of the system was highly reliable. The cable's biggest problems were in the "end terminations" and the "limited bend radius" of the cable.

After the electro-hydraulic system proved to be reliable, more commands to the BOP stack and more monitoring of BOP stack functions were requested by the industry. With the request for more capability, a larger cable than before would be required, with an even greater limited bend radius. This resulted in a trend towards a multiplex or coded system, which did not require the number of wires but required more electronics on the surface and subsea. At the time of this writing there are no field proven multiplex systems, but every control company has its own multiplex system. At present there are four multi-wire electro-hydraulic control systems working in the field.

As deeper water depths are attempted from 2000 feet to 10,000 feet, electro-hydraulic controls are the only way for the industry to go. The main reasons for this are: (1) signal time is reduced to almost zero, (2) more commands are needed on the BOP stack, and (3) more monitoring devices are needed on the BOP and riser.

As these deeper depths are attempted, multiplex offers more advantages to the industry than disadvantages. Those interested in deep water application are encouraged to look at multiplex electro-hydraulic control systems, but the straight hydraulic control system is recommended for water depths of 2000 feet and less.

PART IV

Subsea Production and Diving Operations

16

Subsea Production Systems

W. B. Bleakley
Lockheed Petroleum Services Ltd.

Evolutionary processes in the oil industry are slow since there is a natural resistance to change. But these processes are always at work and certain trends forecast the need or desire for subsea completions for oil and gas wells.

As the potential for onshore oil and gas discoveries diminishes, oil operators and governments are moving offshore. Many countries with shorelines have already granted exploratory permits, and a large number of these have drilling and production concessions (same as U.S. lease) in force.

With this trend worldwide the total offshore acreage under lease has increased rapidly. The Gulf of Mexico sale in October 1974, for example, offered more than 500,000 acres (2,023 square kilometers) to the oil industry.

As the total area under lease increases so does the average water depth. The natural course to follow is to start with near-shore areas and to progress seaward. This seaward progress carries with it the liability of deeper and deeper water. In the Gulf of Mexico sale referred to above, 89 tracts were in 400 to 2,000-foot water depths.

Deepwater Drilling Trends

Over the last few years, the number of floating drilling rigs —both ship-shape and semisubmersible—has increased at a steady pace. Drilling contractors, who are responsible for the design of almost all new rigs, are aware of the trend toward deep water developments and have continually increased the depth capabilities of new design rigs.

As of October 1, 1974, there were 32 drillships and 49 semi-submersible rigs in operation worldwide. At that time, 34 drill-ships and 70 semisubmersible rigs were under construction, and all were due for completion within a 3-year period. Some of these drillships have unlimited depth capabilities because of dynamic positioning and many of the semisubmersibles are able to operate in 3,000 feet of water.

Many oil companies and research groups already have deep water drilling experience, as numerous exploratory wells have been drilled in water deeper than 600 feet. Shell Oil Company has drilled in 2,150 feet of water off the coast of Africa and the Glomar Challenger, a research vessel, has drilled in water well beyond the 2,000 foot mark.

The oil companies have shown confidence in the industry's ability to cope with deep water problems through their heavy financial commitments for deep water leases.

Offshore Production and Reserves

Production data show that 18.9% of the world's oil supply was produced from offshore fields in 1973. This was an increase over the 15% from offshore fields just 2 years earlier. This is even more impressive when realizing that the first offshore oil was produced in the late 1940's. Offshore productive capacity has grown from that meager beginning to more than 10 million barrels/day in 1973. The estimate for productive capacity in 1985 is 25 million barrels/day from offshore fields.

Offshore oil reserves in 1974 were estimated at 137 billion barrels compared to 453 billion barrels onshore. Thus, the offshore area, with 25% of the world's reserves, is responsible for 18.9% of the world's daily production. Estimates for future reserves—those yet to be discovered—show a marked increase for offshore areas. British Petroleum Company's outlook for the future shows ultimate offshore reserves to be 571 billion barrels with onshore reserves to reach 1,038 billion barrels. Thus, while onshore reserves will ultimately double, offshore reserves will increase by a factor of four.

With this well established trend toward deep water development of oil and gas fields, the need for subsea operations becomes clear. This does not imply that all operations will be carried out on the ocean floors, because even the most sophisticated subsea production system requires surface support and facilities. However more and more emphasis will be placed on ocean floor techniques as oil operators move farther from land and into deeper and deeper waters.

Even at this early stage in deep water development, operators have a choice in the approach to take. Subsea production systems are currently available on a commercial basis, and improvements in these systems are almost a daily occurrence. The systems which are in operation now and those which are in advanced stages of development will be discussed separately.

THE SEAL SYSTEMS

Subsea Equipment Associates, Limited (SEAL) is an international consortium owned by British Petroleum, Compagnie Francaise des Petroles, Groupe Deep, Westinghouse Electric, and Mobil Oil Company. The consortium has developed two subsea production systems—the SEAL Single-Well Completion and the Multiple-Well Manifold/Production Station.

The single-well system is designed for individual wells drilled vertically with production rates ranging to 10,000 barrels/day. It is suitable for step-out wells at moderate depths or for complete field development in deep water.

The multiple-well system is intended for use where directional drilling is possible. A number of wells are drilled around the periphery of the manifold/production station through a base template. Wells are connected to the enclosure by a special wellhead connector assembly. The manifold/production station may also be used strictly as a manifold station for fields developed by widely spaced, vertically drilled wells.

The Single-Well System

The single-well system is a remotely monitored and control-led subsea Xmas tree consisting of two assemblies set on a specially constructed base. The lower assembly is called the master valve assembly and the upper one is the production control assembly.

The modular nature of the system combines all shorter-life or highly stressed components into one module—the production control assembly—which can be retrieved by special tools from a surface support vessel.

a) *Fixed Equipment*

This category includes the equipment intended to remain permanently on the ocean floor. Except for original installa-tions, the fixed equipment requires no attention during the lifetime of a well. Figure 4-1 shows an example of a perma-nent drilling base, which is the condition of a wellhead when drilling operations are completed. A special landing base is installed on the drilling base and is sealed perma-nently to the casinghead profile. It forms the lower non-removable portion of a manned intervention system, and provides the mechanical support and sealing surface for the Manned Work Enclosure (MWE) shown in Figure 4-2. This base also provides support and first-order alignment for the internal guide structure as indicated in Figure 4-2. Collec-tively, the fixed equipment incorporates a drilling base, a landing base, a connector, and guide structures.

b) *Master Valve Assembly*

The master valve assembly consists of the lower master valve, an annulus access valve, and a flow-line isolation valve. Presently, it is manually installed and removed using a manned intervention system. The valves can be actuated either manually by operators in the MWE or remotely by means of a handling tool. This assembly provides the means of shutting-in the well and the flow lines for removal of the automated production control assembly.

The master valve assembly is essentially a block of valves for isolating the tubing run and annulus run from the well and the flow line, similar to the lower master valve section of

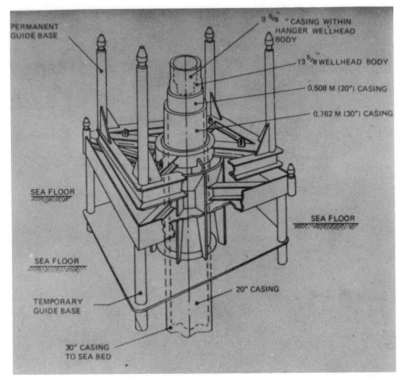

Fig. 4–1   *Permanent drilling base.*

a conventional Xmas tree. It also includes the male half of an electrical wet connector which will transfer power and control to the production control assembly.

The lower master valve package is run suspended inside the MWE (Figure 4-2) which is landed and sealed onto the permanent base previously described. A three-man team is sent down in a personnel transfer chamber which mates onto the control room of the work enclosure. Their job is to winch down, connect, and test the master valve assembly on the drilling wellhead, and to hook up the internal flow line and electrical cable between the base and the master valve package.

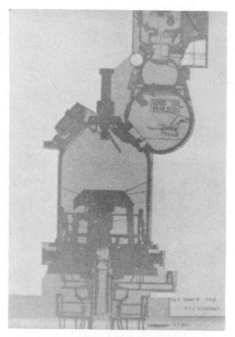

Fig. 4–2   *Manned work enclosure.*

c)   *Production Control Assembly*

This assembly groups together all components which are more subject to wear and will most likely require adjustment, repair, or replacement, such as remote-controlled wing valves, chokes, hydraulic and electronic equipment, sensors, etc. This assembly constitutes the upper portion of the Xmas tree plus auxiliary control systems such as hydraulic, electronic, and electrical distribution subsystems.

The production control assembly can be installed on the seabed, then later removed and replaced (as required) in a sequence of operations performed by a special handling tool without the need for direct human intervention. Once installed, the various functions of the assembly are remotely monitored and controlled by means of a telemetry system connected to a surface control station by a subsea electrical cable.

A prime design requirement was to minimize the overall size and weight of the assembly and the individual capsules it contains. These include the hydraulic power capsule, a downhole safety valve control capsule and hydraulic reservoir, and telecommunications and electrical power. Power requirements, which are low, are supplied from the nearest platform or shore base by a cable, which also carries the telecommunications. The assembly measures about 8 feet in diameter, 10 feet high, weighs approximately 15 tons, and was designed for an operating depth of 1,200 feet.

In line with its design philosophy, the production control assembly was built for a 2-year operating life. At the end of the 2-year working period, the complete assembly is retrieved and a new or refurbished unit is installed in its place. The retrieved assembly is then completely overhauled for use elsewhere.

Maintenance work would not be carried out on the control assembly while it is on the ocean floor. In the event of a component failure, action would depend on the type of failure. If the failure were serious enough to either make the module inoperable or unsafe to operate, the module would be replaced. Figure 4-3 shows the SEAL single wellhead completion and the position of the production control assembly when installed and operating.

d) *Handling-Tool Assembly*

The handling-tool assembly consists of three basic components: the handling tool itself, the reentry system, and the riser. A special handling tool was developed to transport the production control assembly from the surface support unit to the seabed and to install it. During the landing operation, the handling tool provides mechanical protection for the control assembly, absorbs the shocks of landing, and subsequently carries out first order alignment required with the lower master valve assembly.

The reentry system is designed to guide and properly position the handling tool or the manned work enclosure to the subsea wellhead, without the use of guidelines or divers. The reentry system uses sonar to locate the wellhead and position the surface support vessel. Television is used for

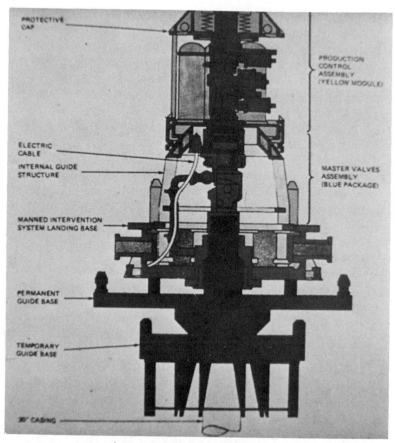

Fig. 4–3 *Final configuration.*

the final approach and reentry itself. This unit is also controlled from the surface.

Finally, the riser consists principally of a drill string with special end fittings, bored to 4-inch ID to pass wireline tools inside, with a reinforced 7-inch section at the bottom. Its principal function is to transport the handling tool to the seabed or to bring it back to the surface and, in the case of human intervention, to handle the manned work enclosure. The riser is suspended from a heave compensation system to absorb the heave when connected to the seabed. It is then used as a conduit to pressure test subsea hydraulic connections, to pump into the well tubing, or to convey wireline tools downhole.

e) *Manned Backup Systems*

SEAL's manned backup systems provide air at atmospheric pressure over the subsea wellhead to enable oilfield specialists to work under simulated surface conditions. These men install the master valve assembly, the flow-line connections within the base, and the electrical cable connections for remote control of the production control assembly.

One system consists of a manned work enclosure and a personnel transfer chamber. The MWE (Figure 4-4) is a pressure vessel lowered to its landing base by a sonar-guided handling tool. The entire assembly is supported by the drill pipe riser from the surface support unit. When landed on the base, the MWE is dewatered and filled with air at atmospheric pressure. Auxiliary to its use is the personnel trans-

Fig. 4-4 *SEAL manned system.*

fer chamber, also shown in Figure 4-4, which transfers the operators from the support vessel to the MWE.

The PTC is an atmospheric diving chamber featuring a buoyancy-control system and a propulsion system. It is linked to the support vessel by an umbilical which provides electrical and hydraulic power. The PTC is lowered and mated to the top of the control room of the MWE, where the men transfer to perform the required tasks. Breathable air and electrical and hydraulic power are supplied to the MWE throughout the operation. After the work is finished, the PTC and MWE are sequentially retrieved to the surface support unit.

During development of the single wellhead completion, it was apparent that there would be a number of subsea production conditions where only occasional well maintenance would be required. For these cases, the SEAL bell (shown in Figure 4-5) was designed as an alternative manned backup system essentially for use with conventional subsea wellheads. The SEAL bell is lowered over the wellhead, then dewatered. Maintenance personnel are carried from the surface in the bell's control room.

After a seal is made from inside the work chamber, men can enter to perform the maintenance required on the subsea completion. This system has been designed to accommodate either conventional subsea Xmas trees, or a specially designed SEAL Xmas tree comprised of a master valve assembly and a simplified production assembly containing essentially only wing valves and power and communication capsules. The SEAL bell system is particularly attractive in depths from 150 to 450 feet. A prototype has been built and tested.

Multiple Well Manifold/Production Station

The station consists of three basic hardware components—the base, the wellhead connector assembly, and the subsea work enclosure. It also includes a number of associated supporting subsystems. A full-scale prototype has been constructed and tested.

Fig. 4–5   *SEAL bell system.*

a) *Base*
     The prototype base (Figure 4-6) is composed of structural
piping and is about 50 feet square, 26 feet high. It weighs
approximately 140 tons in air, with positive buoyancy in
water with pipes unflooded. A circular base was designed to
serve as a template for drilling deviated wells from the sea
floor when the system is projected for use in a production
configuration. For example, a base of 80 feet in diameter was
designed to accommodate drilling up to 18 wells. The actual
size and shape of the base are flexible, provided structural
and operational requirements are satisfied.
b) *Subsea Work Enclosure*
     The subsea work enclosure (SWE) provides a dry envi-
ronment on the sea floor at atmospheric pressure in which
normal oilfield equipment may operate. It is maintained by

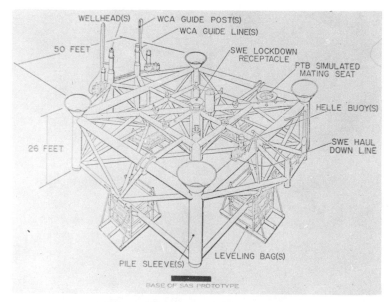

Fig. 4–6   *Base for SEAL system.*

oilfield personnel. The SWE prototype (Figures 4-5, 4-7, and 4-8) is a subsea structure consisting of thick alloy steel measuring 54½ feet high and 16½ feet in base diameter. It is designed for an operating depth of 1,500 feet, weighs 230 tons in air, and is just slightly buoyant in water.

The unit's stepped-cylinder shape was designed to optimize internal piping and equipment arrangement and the external connection of wells completed around its periphery. However, the size and shape of the SWE depend on a number of physical variables, including water depth, number of associated wells, properties of hydrocarbon fluids, production rates, pressures, field characteristics, etc.

c)   *Wellhead Connector Assembly*

This assembly serves as a flow-line link betweed peripheral wellheads and the SWE. A prototype assembly was constructed to connect two wellheads with the SWE through one of its penetrator openings. Early design constraints imposed very wide spatial and azimuthal toler-

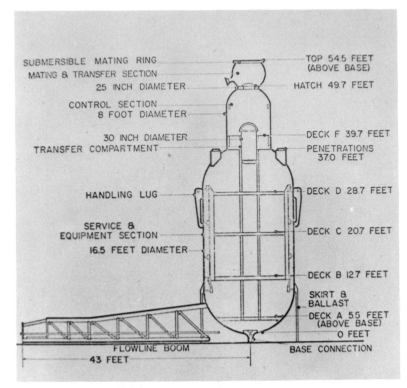

SUBMERSIBLE MATING RING
MATING & TRANSFER SECTION
25 INCH DIAMETER

CONTROL SECTION
8 FOOT DIAMETER

30 INCH DIAMETER
TRANSFER COMPARTMENT

HANDLING LUG

SERVICE &
EQUIPMENT SECTION
16.5 FEET DIAMETER

TOP 54.5 FEET
(ABOVE BASE)

HATCH 49.7 FEET

DECK F 39.7 FEET
PENETRATIONS
37.0 FEET

DECK D 28.7 FEET

DECK C 20.7 FEET

DECK B 12.7 FEET

SKIRT &
BALLAST
DECK A 5.5 FEET
(ABOVE BASE)
0 FEET

FLOWLINE BOOM

43 FEET

BASE CONNECTION

Fig. 4–7   *Subsea work enclosure.*

ances. Thus, the prototype (shown in Figure 4-9) allows for a variation of ± 18 inches between the two wellheads, ± 18 inches between the well couplet and the SWE penetrator, and up to 5 degrees misalignment in either element. It is evident that with today's drilling technology, tolerances of this magnitude are unnecessary and new designs are being initiated which will considerably simplify the connector assembly.

d)   *Supporting Subsystems*

Internally, the SWE prototype consists of an upper control section and a lower service section. The control section houses the electrical equipment and the supervisory control system's remote units. The service section houses all piping

Fig. 4–8 Subsea work enclosure.

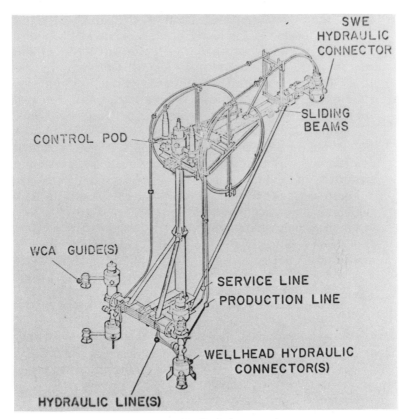

Fig. 4–9 *Wellhead connector.*

equipment involved in producing oil, including a gas-oil separator designed for individual well testing with the capability to automatically run pump-down tools. The prototype also includes life support and environmental control systems, hydraulic and electrical systems, a bilge pumping system, a supervisory control system for automation and monitoring, and a personnel transfer bell for transporting staff to the SWE.

e) *Testing History*

The prototype manifold/production station was dry-land tested in Long Beach, California in 1971, using dead crude oil and gaseous nitrogen to simulate field conditions. The use of the two-phase fluid permitted check-out of the test separator located in the lower portion of the MWE.

In 1972, the unit was transported to the Gulf of Mexico and placed on the ocean floor in 247 feet of water in Main Pass Block 293A near a Sun Oil Company production platform. Production from live wells on the platform was introduced into the system for testing under almost real conditions, including more than 90 automatic operations of the pump-down tool system. During the test period, the personnel transfer bell was used more than 60 times to carry men and materials into the MWE.

Figure 4-10 shows the MWE on a circular base, through which deviated wells are drilled. Figure 4-10A demonstrates the application of the MWE in commingling production from vertically drilled wells completed with the single-well system.

LOCKHEED PETROLEUM SERVICES SYSTEM

The subsea oil and gas production system developed by Lockheed Petroleum Services Ltd. (LPS) consists of three basic parts—the wellhead cellar, the manifold center, and the service capsule.

The system is completely diverless and all components are maintained at one atmosphere of pressure. It is fully automated for remote operation, is capable of through-flow-line (TFL) tool operation, and can be designed for any water depth.

The first system was installed in the Gulf of Mexico in Main Pass Block 290 for Shell Oil Company in the summer of 1972. A reentry was made 1 year later for minor repairs and complete inspection. By January 1, 1975, the system, operating on a dually completed well, had produced more than ½ million barrels of oil. Four additional wellhead cellars and a production manifold were due for installation in the summer of 1975.

The LPS one-atmosphere system has been expanded to provide a means for installing platform riser pipeline tie-ins resulting in a completely welded installation without use of divers or decompression techniques.

Fig. 4–10 *Deep water field.*

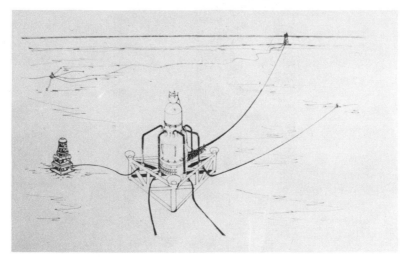

Fig. 4–10A *Deep water field production.*

The Wellhead Cellar

The original LPS wellhead cellar is basically a vertical cylinder (Figure 4-11). It has a lower semiellipsoidal head and an upper hemispherical head penetrated by an upwardly inclined opening called a bullnose port. Flow lines leaving the wellhead or Xmas tree pass through this port. Above the hemispherical

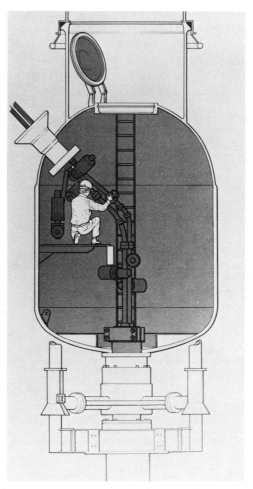

Fig. 4–11 *Lockheed vertical production system.*

head is a short section of vertical cylinder, called a "teacup," which provides a mating surface for the service capsule.

The bullnose port is large enough to accommodate all lines which might be required for operation of the well. In the original installation for Shell Oil Company, the flow-line bundle was made up of two 2⅜-inch diameter lines, one 1¼-inch diameter line, an electrical cable with 34 conductors, and a small-diameter hydraulic hose.

A bullnose to match the bullnose port is used to form a terminal for the flow-line bundle and assist in pulling the lines into the wellhead cellar. A cable attached to the bullnose, with tension coming from a winch inside the wellhead cellar, pulled the bullnose into the port, forming a watertight seal (Figure 4-12). The bullnose is then removed, exposing the pipe ends, and final

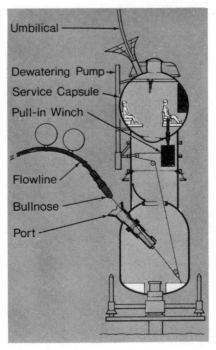

Fig. 4–12   *Flow line, with external TFL loops buoyed to maintain alignment, is pulled into Lockheed wellhead cellar's bullnose port.*

connections are made to the Xmas tree. A preformed loop in the flow-line bundle, just outside the wellhead cellar, provides the necessary 5-foot radius for the TFL tools to pass through the flow line.

a) *Horizontal Wellhead Cellar*

Based on experience with the prototype installation, and looking ahead at a continuing need for TFL capability, the vertical-style wellhead cellar has given way to a new horizontal design (Figure 4-13). In its new configuration, the wellhead cellar provides ample room for crews to perform all operations, and is large enough to accommodate the TFL loops inside.

The inside diameter of the "standard" horizontal wellhead cellar is 10 feet 6 inches, allowing some clearance on each side of the TFL loops with their minimum 5-foot radius. A cellar of this size will enclose a dual 2-inch Xmas tree (two strings of 2-inch tubing plus a 2-inch annulus access) with a pressure rating of 5,000 psi. For larger tubing strings, or for higher pressure ratings, the ID of the wellhead cellar must be increased because of the size of the tree and associated valves.

Wellhead cellars already designed include those capable of withstanding 400, 800, and 1,200-foot water depths in diameters up to 14 feet.

Note in Figure 4-13 that the bullnose port protrudes downward at an angle toward the mudline, facilitating flow-line pull-in.

b) *Installation Procedure*

The wellhead cellar is set in place before the drilling vessel (floater, semisubmersible, jack-up) leaves the location. When drilling is complete, tubing is installed, tubing plugs are in place, and the well is under control, the blowout preventer (BOP) stack is removed.

On the drilling vessel, the wellhead cellar is attached to its guide frame and hydraulic connector. When the well on the ocean floor is ready, the wellhead cellar, with hydraulic connector, is lowered on the guidelines using drill pipe. The hydraulic connector seal is tested and the drilling vessel is released.

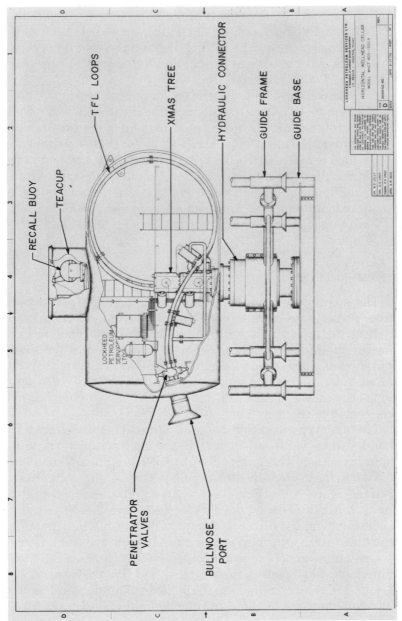

Fig. 4-13   Lockheed horizontal wellhead cellar designed for internal TFL loops.

The heavier components of the Xmas tree, along with the disassembled parts of the TFL loops, are stored in the wellhead cellar during the setting operation so that these large items will not have to be handled in the service capsule.

c) *Role of the Wellhead Cellar*

The wellhead cellar's primary function is to provide protection for the Xmas tree, its associated valves, and the well controls against the hostile environment of the ocean floor. It provides a one-atmosphere habitat for these components which is free of corrosive salt water and high ambient pressure. All of the hardware can be of standard design, off-the-shelf-type equipment. Electrical components are explosion proof, although the wellhead cellar is always inerted (nitrogen gas at one atmosphere pressure) when it is not occupied.

The Manifold Center

The manifold center is designed to commingle production from several subsea wells, each in an LPS wellhead cellar, and deliver the mixed flow stream through a single pipeline to a nearby production platform or shore base. It serves the same purposes as a land-type production manifold as it permits study of an individual well's producing characteristics. It also provides the means to use TFL tools on any of the completions connected to the manifold.

The prototype manifold center (Figure 4-14) is designed to handle three wells, all of which are dual completions with annulus access. It is equipped with 7 bullnose ports of the same design as those on the wellhead cellar. Three of the ports are for producing wells. The flow-line bundle entering each of the three ports consists of two 2-inch flow lines, one 2-inch annulus access line (for gas lift, corrosion or hydrate inhibition, paraffin treating, or other purposes) and one 1¼-inch hydraulic line.

One bullnose port is used for the pipeline carrying the commingled well streams to the platform, and one provides access for the TFL lines coming from the platform. The electrical power and communications cable and the hydraulic fluid supply line enter through a common port. The seventh port is a spare one.

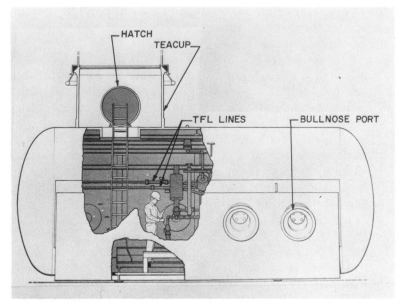

Fig. 4-14   *Three-well manifold.*

Although all of the functions of the manifold center could be carried out on the production platform its presence reduces the number and length of flow lines needed on the ocean floor. For example, a manifold center capable of handling 16 wells is currently in the conceptual design stage. It will commingle production from 16 wells and provide TFL capability from the platform deck.

Without the manifold center, and assuming 2-inch lines to be large enough, 48 separate 2-inch lines would have to be laid on the ocean floor. Two flow lines, one annulus access line, a hydraulic fluid supply line, and an electrical cable would be required for each well. The expense involved in this method of operation is beyond comprehension. Any effort to anchor a vessel of any type in the area would surely meet with disaster.

With a manifold center, flow lines from each wellhead cellar are of minimum length and only one flow-line bundle must be run from the manifold center to the platform.

The manifold center is equipped with a teacup of the same

dimensions as one on the wellhead cellar to provide a mating surface for the service capsule.

a) *Template Concept*

In cases where it is feasible or desirable to use directionally drilled wells, templates are available to support several wellhead cellars. One is designed for installations requiring TFL capability and the other is for non-TFL operations.

Figure 4-15 shows the template and wellhead cellar arrangement for a non-TFL system, incorporating the manifold center at one end. This template provides two hydraulic connectors—one for the well and the other to connect the wellhead cellar's output to the fixed template piping leading to the manifold.

The manifold performs all functions expected of a land-based manifold, but it is unable to handle TFL tools. Wells can be put on or off production, each can be tested, chokes can be adjusted, and the flow line pigged, all by remote control.

If TFL capability is required, the template is used to support horizontal wellhead cellars as described earlier and the

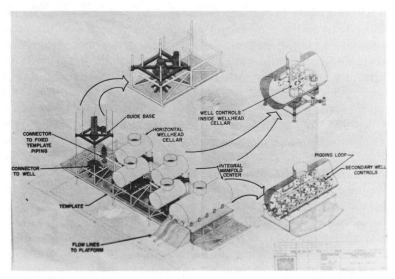

Fig. 4-15   *Template with built-in manifold for non-TFL applications.*

manifold center is installed separately. In a typical installation two templates would be used, each having a capacity for eight wells. They would be located a distance apart as determined by recommended well spacing. The manifold center, designed for 16 wells, would be located conveniently between the templates with separate flow lines to each of the wellhead cellars. A single flow line would then carry the production from the manifold center to the production platform or to shore.

The advantages of this system are that all single-well flow lines are approximately the same length and all are laid in the same "corridor." This simplifies installation and minimizes damage by localizing the "danger zone."

The Service Capsule

The connecting link between the ocean's surface and the equipment on the ocean floor is the service capsule (Figure 4-16). It provides the means for getting men and materials to the wellhead cellars and manifold centers.

The service capsule is basically a sphere with a hatch above and below and a skirt on the bottom. The spherical portion is maintained at one atmosphere of pressure during all subsea operations, but the skirt is flooded while the capsule is ascending or descending.

The capsule is positively buoyant by about 2,000 lbs. and must be winched down to mate with the wellhead cellar or manifold center. A recall buoy in the wellhead cellar's teacup can be released acoustically or electrically. It carries a messenger line to the surface. When surfaced, the buoy is retrieved and the messenger line is used to pull up the haul-down cable. The cable is then fastened to the winch in the skirt of the service capsule and the capsule winches itself down to the wellhead cellar.

At the cellar, the flange of the skirt mates with the top ring of the teacup and the skirt's gasket is compressed. At this point, a new volume is formed, but it is full of water which resists compression. The capsule dewatering pumps are started and the trapped water is pumped overboard. Once this process starts, the capsule is seated on the teacup's mating surface and is held

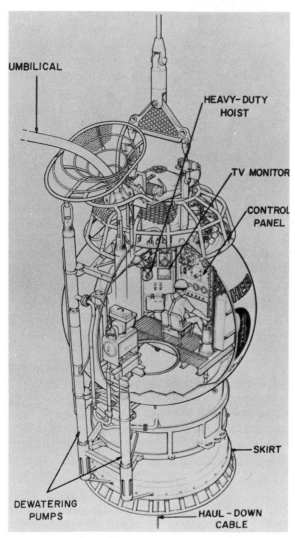

UMBILICAL

HEAVY-DUTY HOIST

TV MONITOR

CONTROL PANEL

SKIRT

DEWATERING PUMPS

HAUL-DOWN CABLE

Fig. 4–16 *Lockheed's service capsule.*

there under the full hydrostatic head of seawater. Dewatering continues until the newly formed volume is dry, at which time the capsule's lower hatch can be opened.

The atmosphere in the wellhead cellar is then tested to determine its composition and to insure that no hydrocarbons are present. The cellar is purged to remove the nitrogen left in the cellar at the last visit and to replace it with air at one atmosphere of pressure.

When the cellar's atmosphere is safe, its hatch is opened, allowing LPS crewmen to enter for routine maintenance, repair, or inspection. No special training is required for this mission, no diving apparatus is needed, and there is no wasted time for decompression. All operations are carried out in a shirt-sleeve environment, with conventional oilfield tools handled by men with land-type oilfield training.

When work in the wellhead cellar is completed, or when it is time for a crew change, the entrance procedure described above is reversed. Crewmen leave the cellar and close its hatch, move into the service capsule and close its hatch, then equalize the pressure in the teacup area with that on the ocean floor. The capsule then ascends as the winch pays out the haul-down cable.

If all work is completed in the wellhead cellar and another visit is not scheduled immediately, two major changes are made in the departure sequence. First, a new recall package is installed in the teacup so that a haul-down cable will be available for the next visit. Second, the wellhead cellar is purged of oxygen and left with a nitrogen atmosphere.

Surface Support Vessel

Work on the ocean floor would be impossible without support from the surface. The LPS surface vessel (Figure 4-17) provides the power, life support, and communications needed to safely operate far beneath the sea.

One of the support vessel's prime functions is to launch and recover the service capsule. Since the capsule is buoyant, it operates independently once it is in the water, using its built-in winch and the haul-down cable. When surfaced, however, it

Fig. 4–17   *Lockheed's surface equipment.*

must be picked up out of the water and stowed on the surface support vessel's deck. It must be placed in the water at the time of the next dive. A complicated hoisting system accomplishes these procedures for the LPS service capsule.

Once beneath the ocean surface, the service capsule is supplied with air, electricity, and communications lines through the umbilical. A life-support van on the support vessel provides conditioned air to the capsule so temperature and humidity are under control at all times. The air source maintains capsule pressure just above normal atmospheric pressure. Fresh air enters through one conduit in the umbilical and exhaust air leaves through another conduit of the same size.

Power cables in the umbilical provide the capsule with 440-volt, three-phase electricity for lighting, hydraulic pump power, and other needs.

The umbilical also carries a number of communications lines to permit voice and video contact between the capsule and the surface and to transmit readouts from various instruments. The support vessel monitors capsule pressure, oxygen level, carbon dioxide content, and presence of hydrocarbons. An unfavorable reading of any of these quantities sounds an alarm. Visual read-outs of these quantities are available in the capsule and in the control room on the surface.

The capsule carries four TV cameras—one fixed inside, one fixed in the skirt, one outside for viewing flow-line operations, and a portable camera for use as needed. The control panel in the capsule has a TV monitor which can be switched to display the picture from any of the cameras.

Two TV monitors on the support vessel have the same capability as monitors on the capsule's control panel. Any two of the four cameras can be monitored at one time. Generally, the fixed camera in the capsule will be tuned to one of the surface monitors so the support crew can watch the capsule's crew. Also, the capsule's crew and surface control room personnel are in continuous voice contact.

Flow-line Pull-in

The unique LPS diverless flow-line pull-in system is a significant part of the entire program as flow-line installation is usually expensive and subsea connections normally require diver assistance.

As pointed out earlier, the original flow-line connection for Shell Oil Company was accomplished using a preformed line with external TFL loops pulled by a winch within the service capsule. With the horizontal wellhead cellar design, the flow line approaches the cellar from the ocean floor and is lifted at a slight angle to align the flow-line bundle's bullnose with the cellar's bullnose port.

A new pulling system has also been adopted to permit use of larger forces with the reaction directly on the hull of the wellhead cellar or manifold center.

The dry pull-in system (Figure 4-18) uses a linear device which operates in a hand-over-hand manner to haul in the pulling cable with 10-inch strokes. The cable passes through a wireline stripper and a line wiper to prevent entry of seawater

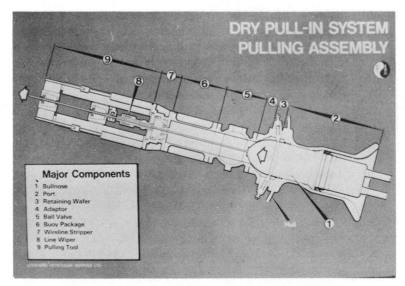

Fig. 4-18   *Dry pull-in system pulling assembly.*

during the pulling operation. As shown in Figure 4-18, the entire pulling assembly is connected to the inside of the bull-nose port. All can be removed once the bullnose is seated and sealed.

At the beginning of the pull-in sequence, a buoy is located in the buoy package (shown in Figure 4-18). The ball valve is opened and the buoy is released to the surface, carrying with it a messenger line. The buoy is retrieved on the surface and the messenger line is used to pull the pull-in cable to the surface. There it is attached to the flow-line bullnose on the pipelay barge. The barge lowers the flow line into the water as additional pipe joints are added, and the cable is simultaneously pulled into the wellhead cellar.

When the bullnose is pulled into the bullnose port, it is locked in place in the retaining wafer (shown in Figure 4-18) and the seal is tested. With the seal seated, the puller assembly is removed. This exposes the flow-line bundle bullnose which is then removed, permitting hookup of the flow lines to the Xmas tree or the manifold.

Application to Platform Risers

The LPS dry flow-line pull-in technique has been adapted to large-diameter pipelines used to move oil from a production platform. In this application, the platform riser pipe is equipped with a connector chamber, or pressure hull, having a top flange which will mate with the LPS service capsule (Figure 4-19A). The riser pipe penetrates the wall of the pressure hull and is welded to a larger-diameter pipe section. This in turn is flanged to the bullnose penetrator.

At the time of installation, a buoy is located in the bullnose port and is connected to a line running back through the riser pipe to the platform deck. This is the pilot line which will be used later to draw the heavier pull-in cable through the riser and back to the pipelay barge on the surface.

Figure 4-19, A-D shows the sequence of events followed in the completion of the tie-in. With the pipelay barge in place and the pipe end equipped with a bullnose, the buoy is released and retrieved at the surface. The pilot line on the buoy is used to haul the pull-in cable from the platform deck down the riser, through the pressure hull and bullnose port, and to the surface. There it is joined to the pipeline's bullnose and the pipeline is lowered into the water.

Using platform-deck power, tension is applied to the pull-in line and the bullnose approaches the bullnose port (Figure 4-19B). As pulling continues, the bullnose enters the bullnose port where seals are engaged to make a watertight joint. If additional pull-in force is required, the riser pipe can be dewatered by lowering a submersible pump. This creates a differential across the bullnose due to the hydrostatic head of seawater, driving the bullnose home and affecting a seal. At this point, the pipelay barge is free to continue pipelaying if it is laying away from the platform.

The Lockheed service capsule can now be winched down to mate with the pressure hull (Figure 4-19C) or this operation can be delayed to a more convenient time. When in position, the hull is dewatered to form a one-atmosphere work chamber.

Using a cutting tool powered by the capsule's hydraulic system, a short section of riser pipe is cut away, and the larger-

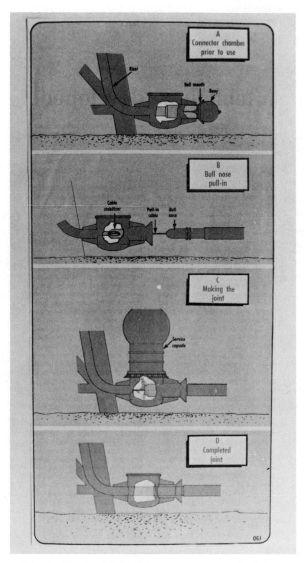

Fig. 4–19 *Lockheed's sequence of operations.*

diameter transition pipe section is unflanged and moved clear. The bullnose can be welded directly to the bullnose port at this time, if desired.

The cable is removed from the bullnose and the bullnose is cut off. A precut and beveled pup joint is lowered into place and welded, connecting the pulled-in pipeline to the platform riser (Figure 4-19D). The joint is X-rayed and coated for corrosion protection.

Welding under these conditions presents no problems as the capsule air supply is sufficient to remove welding fumes to eliminate danger to the operator. Further, the riser pipe is open to the atmosphere, helping to remove those vapors that gain access to the inside of the pipe.

EXXON'S SUBMERGED PRODUCTION SYSTEM (SPS)

Exxon Company U.S.A. has developed a submerged production system (SPS) which has undergone extensive land testing. It is currently in the submerged test phase under 170 feet of water in the Gulf of Mexico 27 miles southeast of Grand Isle, Louisiana.

The SPS consists of a massive template which supports a number of subsystems. Each subsystem is designed to perform a given function in the production of oil and gas from offshore fields. The system is capable of deep water operation and can be maintained and repaired without the need for divers. A remotely controlled manipulator removes and replaces components as needed, but will carry a man to the ocean floor in a shirt-sleeve environment if necessary.

Many conventional oilfield components have given way to specially designed equipment items which insure long life and functional reliability. Dry-land tests on carefully selected components lasting as long as 1,000 hours bolster Exxon's confidence in the SPS.

The Manifold

This unit (Figure 4-20) is the heart of the entire system. It is

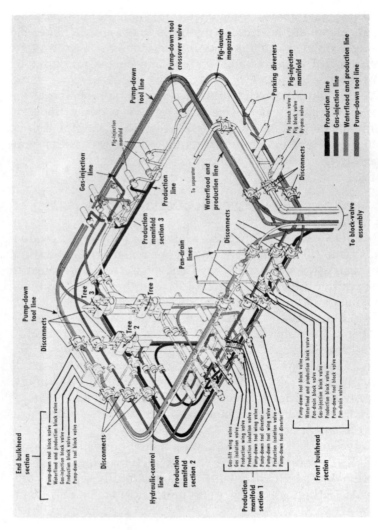

Fig. 4–20 *Production manifold for Exxon SPS.*

made up of all the necessary piping to carry out the assignments of the SPS. It is rectangular shaped with well-rounded corners to allow use of pump-down tools. In the prototype, its shape is almost square because the design is based on three wells. But the square can be elongated to a rectangle by adding manifold sections for additional wells.

The manifold is fail-safe. If hydraulic power is lost, or if any abnormal conditions exist, all valves close and the system is left in a safe condition.

The manifold is designed to permit use of pump-down tools, to provide for individual well tests, to handle injection of chemicals for inhibition of corrosion or hydrate formation, and to provide for artificial lift and control producing rates through remotely adjustable chokes.

The manifold is actually a dual system which permits injection of pressure-maintenance fluids through one group of wells while simultaneously producing another group. Further, the manifold has the capability of pigging all lines in the system.

To achieve all these functions, the manifold consists of 5 lines. Two large lines can be used for high and low-pressure gathering systems, or one can be used for production and the other for injection. Two other lines are multipurpose lines that can be used for inhibitor injection, well testing, and pump-down tools. The fifth line is for gas-lift gas distribution.

Each well is connected to all of these lines. This makes the system relatively complex but eliminates addition of new equipment as producing conditions change. The manifold is ready to handle production from the time the well starts flowing, through the artificial-lift period, and on to abandonment. The manifold can handle pressure-maintenance chores as needed and perform well repair and maintenance via the pump-down system. It is designed for initial installation with all these capabilities so nothing need be added at a later date.

Only two kinds of joints are in the manifold—welded joints and mechanically locked metal-to-metal seals. They provide long life and good reliability.

To insure safety in the manifold, it was designed according to the most stringent code requirements. This was done to take advantage of all available technology in metallurgy, design for high stress levels, and fabrication procedures. Exxon engineers felt that a rigid code would include the toughest requirements in these three areas.

The multipurpose lines are rated at 5,000 psi, the gas-lift lines at 1,500 psi, and the large gathering manifold at 3,000 psi. On completion of fabrication, these lines will be tested at twice these rated pressures. All welds are 100% X-rayed.

The Wells

Wells drilled to produce through the SPS are rigged for pump-down tools, using strings of 3½-inch tubing. Each well is equipped with a full-opening, surface-controlled, hydraulically operated down-hole safety valve, located below the mudline. The valves are fail-safe and are held open by hydraulic pressure.

At the Xmas tree, there is a conventional fail-safe master valve which passes produced fluids through looped lines to the manifold. The looped lines permit passage of the pump-down tools.

The trees are more simply constructed than conventional trees in that the wing valves and other appurtenances have been moved to the manifold. Collet connectors join the wellhead to the manifold. A specially designed tool applies hydraulic pressure to the mating part of the connection to complete the joint.

Well Completion

The wells are designed to permit drilling, landing the tree, and completion using a conventional floating drilling rig. Tubingless completions are a part of the prototype SPS, as this design fits the requirements of the system. The wells will have the capability of vertical reentry workover if necessary.

Control

Control of valves and other parts of the SPS is through a closed hydraulic system as a basic power source. The pressure in the system is subsea-generated with all equipment mounted on a replaceable skid.

The system is designed so there is a hydraulic sump on the

low-pressure side of every packing. Thus, if there is any leakage, it is from a high-pressure to a low-pressure hydraulic system. There are no seals with high-pressure hydraulics on one side and seawater on the other. The low-pressure system is maintained at a slight amount above sea pressure. This pressure is maintained by equilibrating hydraulic pressure with seawater pressure, but at a lower elevation than any point in the hydraulic system. This difference in hydraulic head from the two fluids creates the difference needed to prevent seawater contamination of the hydraulic system.

The hydraulic pump is electrically driven (3 hp) through a submarine power cable. This multiconductor cable also carries the communications circuits. Hydraulic-pressure generation is automatic, but may be controlled from the surface to maintain 3,000 psi in the system.

A gas-type accumulator in the system provides sufficient energy to continue routine operations, opening and closing of valves, etc. for a 7-day period in case of power failure. This allows sufficient time for repairs to be made before the fail-safe system automatically shuts in the production system.

Pump and Separator Subsystem

This subsystem is designed to input energy into the produced liquid stream to improve flow rate and ultimate recovery from the reservoir. Figure 4-21 shows a model of this unit. The separator is a gas-liquid separator and no effort is made to separate the produced water from the oil in this process. The produced liquids provide a flooded suction for a bank of multi-staged centrifugal pumps.

The pump discharge flows to the surface processing facility with a portion of the stream returned to the separator to maintain the liquid level. Pumps are automatically turned on or off to meet the long-term variations of flow rate. Short-term variations are handled by a throttling valve which automatically regulates the portion of the pump's discharge returned to the separator. Hence, the pumps on stream must have a greater capacity than the production rate. The liquid level in the separator vessel which is used to control the throttling valve and the number of

Fig. 4–21   *Model of subsea gas-oil separator and pump station.*

pumps operating is monitored by radioactive means. An electrohydraulic control system is used to control the unit.

Template

The template (Figure 4-22) is the large tubular structure used to carry all of the subsea equipment to the seabed and to provide the alignment for connecting the wells and pipelines to the manifold. The template is designed to permit the subsea equipment to be preassembled and debugged on land before installation and then to be installed as a unit. The structure has sufficient strength and displacement to contain the preassembled equipment while being launched from a large barge, being selectively flooded to negative buoyancy and lowered to the seabed by a drilling barge.

The template subsystem required essentially no new components. Hence, little land testing was required to ready it for offshore testing. Slips which engage a pile and are needed for leveling have been reduced to prototype hardware and tested at 400 tons. Special handling tools for setting the piles, leveling

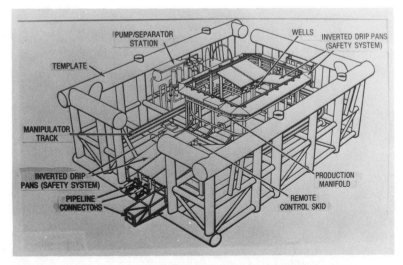

Fig. 4–22   *Template for Exxon's subsea production systems.*

the template, and releasing the lowering bridles have also been built and tested.

Maintenance Manipulator

The maintenance manipulator subsystem (Figure 4-23) is a set of equipment and procedures for maintaining the manifold, remote control, and pump and separator subsystems. A workover rig is required only for maintenance of the down-hole safety valves and the tree, for vertical reentry workovers on the well, and for replacing the fail-safe pipeline valves. The manipulator has an air weight of approximately 68,000 lbs., but it floats with about 1,500 lbs. of positive buoyancy in seawater.

This large machine is the primary work tool rather than just being a rigger that rigs tools to permit work to be done. Maintenance is by replacement of the maintenance module which contains the malfunctioning component. The manipulator brings down a new piece, removes the old piece, inserts and pressure tests the new piece, and returns to the surface with the old piece all in one deployment. The subsea equipment, though

Fig. 4–23   *Exxon's maintenance system.*

designed to minimize maintenance requirements, is built to permit replacement of all underwater components except the template and wellheads.

The prototype manipulator has been thoroughly tested and is operational. All maintenance positions on the manifold have been achieved. All valves and actuators have been repeatedly removed and reinstalled. The special handling tool for that operation delivers approximately 15,000 ft.-lbs. of torque to preload the sealing surface of the valves. All electrohydraulic control pods have been removed and installed.

The special handling tool for this operation handles loads up to 1¼ tons with the gentle precision required to stab a multiported flange. These activities are controlled via a television camera placed to provide the best vantage point for observation. The critical alignments are about 2 feet from the camera lens, thus minimizing the visibility problems of turbid water. Trial deployment of the manipulator in 425 feet of water has successfully demonstrated the ability of the manipulator to land and latch on to the track encircling the manifold.

Test Procedure

To complete the test phase now underway in the Gulf of Mexico, Exxon drilled three directional wells into a proven reservoir. The prototype unit was designed for only three wells, but the system can be expanded to as many as 40 wells according to ·Exxon engineers.

The subsea test program culminates more than 300 man-years of research and development and represents an expenditure of more than $30 million. The overall program, visualized in Figure 4-24, involves tests of all the subsystems as well as a riser system and monobuoy. Production can be either off-loaded at the buoy, simulating deep water operation, or be moved to a nearby production platform.

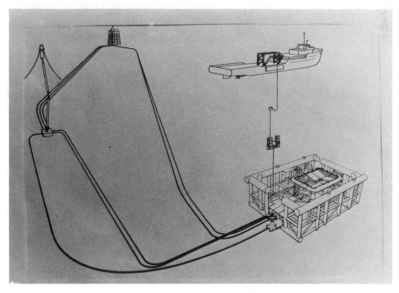

Fig. 4–24 *Exxon's system schematic.*

REFERENCES

1. Chatas, A. T., "SEAL Subsea Petroleum Production Systems," *Proceedings,* Tenth Annual Conference, Marine Technology Society, p. 529.
2. Adams, J. R., Manual, J. R., and Cook, R. W., "New Offshore Recovery Advancements for Deep Water Completions," Preprints, Offshore Technology Conference, Vol. II, No. 1834, May 1, 1973, pp. 133-148.
3. Bleakley, W. B., "Production Unit Ready for Subsea Test," *Oil and Gas Journal,* August 9, 1971, pp. 55-61.
4. Chatas, A. T., and Richardson, E. M., "Subsea Manifold System," Preprints, Offshore Technology Conference, Vol. I, No. 1967, May 6-8, 1974, pp. 341-350.
5. de Panafieu, P., and Darnborough, E., "The SEAL Intermediate System," Offshore Scotland Conference Paper, *Offshore Services* Magazine, March 1973.
6. Penncock, M. D., "Application of Subsea Production Systems to the North Sea," Presented at Technology Offshore (North Sea) Conference, Paper C, 1971, Multidere Ltd.
7. Walker, R. I., and Ayling, L. J., "Manned and Remote Intervention for Subsea Oil and Gas Production," Presented at Petroleum and Mechanical Engineering Conference, ASME, September 15, 1973.
8. Bleakley, W. B., "Lockheed's New Underwater Wellhead System Ready To Go," *Oil and Gas Journal,* Aug. 7, 1972, p. 67.
9. Bleakley, W. B., "Lockheed, Shell Flange Up Subsea Well," *Oil and Gas Journal,* Oct. 23, 1973, p. 64.
10. Bleakley, W. B., "Shell Reenters Subsea Wellhead Chamber," *Oil and Gas Journal,* July 29, 1974, p. 158.
11. DeJong, J., and Brown, R. J., "Development and Utilization of a Deepwater Pipeline Connector," Preprints, Offshore Technology Conference, No. 1835, April 29 - May 2, 1973.
12. DeJong, J., "New Riser Tie-in Technique Developed," *Oil and Gas Journal,* May 28, 1973, p. 49.
13. Scott, R. W., "How Humble Plans to Produce Oil in the Santa Barbara Channel," *World Oil,* Dec. 1970, p. 39.
14. Bleakley, W. B., "Humble's Subsea Production System Nears Test Phase," *Oil and Gas Journal,* Aug. 30, 1971, p. 49.
15. Burkhardt, J. A., "Test of Submerged Production System," SPE Paper No. 4623, presented at 48th Annual Fall Meeting of SPE, Las Vegas, Nevada, Sept. 30 - Oct. 3, 1973.
16. O'Brien, D. E., "Exxon System Slated for Test," *Ocean Industry,* July 1974, p. 15.

17

Diving Operations and Equipment

Mike Hughes and
Larry Cushman
Oceaneering International, Inc.

Diving can be defined as a method of transportation that allows a man to perform work in the ocean, or on the ocean floor. There are two basic ways men can work underwater. The first is pressurized (or hyperbaric) diving. With this method the man goes directly into the ocean, and is therefore continuously subjected to the increasing pressures of depth. The second method of diving is unpressurized (or isobaric) diving in which the man goes under water in an unpressurized sealed diving bell, submarine, or other device that is capable of withstanding external depth pressure.

The advantage of an unpressurized diving system, such as a submarine, is that the occupant is protected from the conditions inherent in a high-pressure, underwater environment. He is not subject to the physiological problems caused by increased pressure, such as oxygen toxicity, carbon dioxide poisoning, inert gas narcosis or decompression sickness. However, these unpressurized systems also protect the environment from the man; that is, a man enclosed in a protective submersible diving system cannot perform work as effectively as a man actually in the underwater environment.

Today, diving technology is sufficiently advanced to permit the most effective underwater work to be performed by divers in

the water, in spite of the physiological and psychological problems that result from hyperbaric exposure.

History of Diving

The story of hyperbaric diving begins with the first pre-
historic breath-hold divers. The first recorded event took place
in 415 B.C., when Greek warrior-divers destroyed a harbor boom
at Syracuse. History indicates that open-bottom diving bells were
also in use in the fourth century B.C. Interest in diving began to
increase in the 1500's when diving systems of different types
and of varying degrees of practicality were designed and constructed. Some were used to carry out surprisingly complex
underwater operations, usually for ship salvage and cargo recovery.

Diving as we think of it today began in 1819 with the development of the "open-helmet" system by Augustus Siebe, in
England. This equipment, with minor modifications (for example, the addition of the full enclosed suit in 1837) is the basic
"hard-hat" or deep sea diving system which is the forerunner of
all present day diving equipment. Compressed air is pumped
down from the surface into a metal helmet in which the diver
breathes. The helmet can be thought of as a small open-bottom
diving bell.

The pressure of the compressed air in the helmet is always
equal to the surrounding water pressure because the excess air
(pressure) escapes at the bottom, and the diver can comfortably
equalize this pressure and breathe underwater without difficulty. This diving system was not used earlier because compressed air was not available. Truly, the capability to mechanically compress large volumes of gas made underwater diving
possible.

In 1906, Dr. J. S. Haldane in England made an important
contribution to diving when he developed a firm mathematical
and physiological basis for solving the problem of decompression. Dr. Haldane developed the stage method for decompression, and modified Haldane methods are used today for comput-
ing decompression schedules.

During World War II significant changes were made in development of diving equipment. Oxygen rebreathers (closed-circuit breathing systems) were designed which would allow men to operate underwater without being detected by their exhaust bubbles, and more elaborate diving equipment was developed for salvage operations, submarine rescue and other underwater work. Compressed air was available at pressures high enough to allow a man to comfortably carry a practical air supply in cylinders on his back, and self-contained air diving equipment (scuba) became a standard diving technique.

In the early 1960's most of the world became intrigued with the idea of living and working underwater. During this decade some important underwater experiments were conducted. The United States and France carried out underwater habitat experiments such as the "Man-in-Sea" project (Ed Link), "SEALAB" (U.S. Navy) and "CONSHELF" (Cousteau-France). These projects tested the new concept of "saturation" diving, or long-term living under pressure down to depths of more than 300 feet. The success and glamour of these programs created an interest in underwater technology that resulted in the growth of many other ocean-related activities, such as the development of complex submersible vehicles, underwater life support systems, and new types of diving equipment.

In 1967 a number of commercial diving companies recognized the immediate need for deep diving services for the rapidly expanding offshore petroleum industry. Open water dives to depths of more than 600 feet were conducted to demonstrate that the necessary commercial diving services would be available when the oil companies started drilling in deep water.

However, the "age of underwater experimentation" began to slump in the late 1960's when it became clear that research diving programs were not going to result in immediate economic returns. Much of the scientific and industrial experimentation was suspended, and most of the more than 70 submersible vehicles built by aerospace companies were put into storage or sold. And yet the practical development activities in oil-related diving continued to increase. The commercial diving companies that demonstrated their ability to perform useful work underwater in the field became the leaders in the development of advanced diving procedures and equipment.

OIL FIELD DIVING—
THE STATE OF THE ART

Although most offshore drilling is taking place within the 600-foot water depth range, the offshore petroleum industry has set its sights on deeper waters. Several diving contracts calling for 1000-foot water depth capability have already been awarded, and requests for diving services beyond 1000 feet are becoming common.

It is not unreasonable to assume that 2000-foot diving services will be needed in the near future, and deeper services will ultimately be required to keep pace with the depth capabilities of offshore drilling equipment presently in the concept-design stages.

The Equipment Problem

The subsea drilling and production systems used by the offshore petroleum industry have been designed as "diverless" systems. However, provisions are usually made so that divers could be used in case of emergency. A great deal of effective work has been done in depths up to 300 feet, and operational 600-foot dives from drilling vessels are considered well within the state of the art today. Many of the bell diving systems presently in use are 600-foot systems.

Successful subsea drilling operations have been conducted in water depths of 1300 feet. One of the problems facing the diving industry, however, is that the design criteria used for 600-foot diving systems will not be adequate for deep diving complexes. In general, 600-foot systems were redesigned from 300-foot systems simply by increasing the vessel and plumbing pressure ratings and changing out the gauges and regulators. This type of "redesign" is no longer adequate. The new deep systems must be designed specifically to solve the critical problems that result from increased depth, such as precision gas control and monitoring, protection from extreme cold water exposure, and diver comfort in the deck chambers during the long decompression schedules.

The Narcosis Problem

All diving may be broadly categorized into either "air diving" or "mixed-gas diving." The former, in which the diver breathes air, has physiological limits dictated mainly by the narcotic effects of nitrogen. An optimum depth of about 180 feet is generally accepted throughout the industry for working dives using air. Below this depth, part or all of the nitrogen must be replaced by another inert gas, usually helium, which has no apparent narcotic effect on the diver.

The Decompression Problem

A diver must decompress from depth according to a predetermined schedule so that the inert gases (i.e., nitrogen and/or helium) which are absorbed in the body under pressure may be released from the body tissues without the bubble formation which would cause decompression sickness or "bends." The quantity of inert gas absorbed during a dive is a function of pressure (or depth) and duration of exposure (or bottom time).

The total decompression time varies with the quantity of absorbed gases and is therefore also a function of depth and bottom time. There are various techniques used to minimize total decompression times so that available equipment and personnel can be used to maximum efficiency without compromising safety considerations. The most common technique is "short-duration" diving.

Short-Duration Diving

The term "short-duration" or "bounce" diving (sometimes called "nonsaturated" diving) simply means that a man is pressurized for a relatively short period of time (usually less than one hour) and then decompressed in a matter of hours. Due to the short exposure to the increased pressure gradient, the diver's body has absorbed a relatively small amount of gas and his decompression can usually be satisfactorily completed in less than 24 hours. The term "nonsaturated" is really technically incorrect. Some portions of the diver's body (i.e., his bloodstream) are saturated very quickly after he reaches a stabilized pressure. Because they saturate quickly, these portions also desaturate (decompress) quickly.

Today's physiological knowledge is based upon an extensive history of how long it takes to decompress these "fast" tissues. A reasonable limit for today's conventional "short duration" diving capability is a half hour at a maximum of 600 feet. There has been a lot of talk, but not much diving, beyond that depth. All decompression tables are based on theoretical calculations which have been followed by simulated dives. If these tests prove adequate, the tables are tested under field operating conditions and modifications are made based upon the results. There simply has not been enough operational use of tables beyond the 600-foot level to provide the empirical data necessary to optimize the decompression schedules and minimize the incidence of bends.

Short-duration diving is a very important part of drilling rig support operations. Divers working on drilling rigs have known for some time that the vast majority of drilling rig problems can be solved with one or two very short dives. An analytical study of over 3000 drilling rig working dives on drilling rigs was made by Oceaneering International which substantiates this. The study showed that more than 80% of all working dives were shorter than a half hour, and 99% were completed in less than one hour. A similar study completed for a drillship operating in the Irish Sea during 1971-72 showed a total of 222 dives averaging 16 minutes per dive.

The important factor is that most of the diving requirements onboard drilling rigs can usually be satisfied with dives of less than a one hour duration. Virtually all could be satisfied with two dives of a half hour each.

Saturation Diving

The concept of saturation diving was introduced to the offshore oil industry in 1966. It was originally conceived to provide a more reasonable ratio of bottom working time to decompression time. For any given depth there is a maximum time beyond which the inert gas effectively saturates the body tissues. Decompression time becomes a function of depth only. Saturation diving differs from short-duration by the fact that there are no decompression penalties for greatly extended bottom times. This advantage can be offset by depth, however. Decompression schedules can take a week or more to return a diver to atmospheric pressure from very deep dives. A rule of thumb for estimating saturation decompression is one day of decompression time for each 100 feet of depth.

Theoretically, the body does not become completely saturated at any pressure until the diver has remained there for about 24 hours. In practice, however, saturation decompression is required any time the diver has stayed at a depth for a time beyond which he can be decompressed using conventional "short duration dive" decompression.

Why can't we use conventional decompression for a man who has remained at 600 feet for two hours? He is obviously not totally saturated in that length of time. The answer is simply that neither theoretical nor tested information exists. Everyone knows that a man could be decompressed much faster than saturation would require (about 6 days), but no one knows how to do it safely. Even those who think they know how to have never conducted the field tests to prove that such a technique is safe. The only resort, then, is saturation decompression.

The Compression Barrier

Even if adequate proven decompression tables existed, diving deeper than 600 feet is further complicated by what can be described as the "compression barrier." A diver's body begins to absorb the inert gas he is breathing as soon as the pressure starts to increase. Therefore, his bottom time is considered to start when pressurization begins. Since his dive is limited to a specific total time, any reduction in compression time leaves more time for useful work on the bottom.

We know from experience that men can be compressed to a depth of about 600 feet in a matter of minutes (usually ten or less). If our limitation is a half hour at 600 feet, then the diver has at least 20 minutes of useful bottom time before beginning decompression.

Tests have also shown that rapid compression to depths greater than about 600 feet can cause a serious problem known as High Pressure Nervous Syndrome (HPNS). The HPNS mechanism is not well understood, but its effects are severe —drowsiness, dizziness, tremors, and nausea. This leaves us with a diver on bottom who is not only incapable of working, but who could become a hazard to himself. If we cannot compress divers rapidly, by the time they reach the bottom they will already have incurred a substantial amount of time under pressure and, therefore, a very lengthy decompression obligation.

The present solution for preventing HPNS is to compress the men very slowly (40 feet per hour according to the U.S. Navy tables) and then give them saturation decompression at the end of the dive. In practical terms, this would require 25 hours to compress down to a depth of 1000 feet. For just one hour of bottom time, the diver would be committed to 25 hours of compression, one hour on bottom and ten days of decompression. This is an unacceptable price to pay for an hour of useful work.

Much of the concern in the commercial diving community today centers around the need for relatively short bottom times at depths greater than 600 feet. This will probably be a requirement for future drilling operation. If a way can be found to pressurize the men safely to the bottom depth within a matter of minutes, it is possible that relatively short decompression times could be used following work periods of a half hour to one hour. To do so safely, however, will require a great deal of research and field testing.

The Crossover Point

Diving to 1000 feet is not one of the industry's major problems at this time. Our biggest concern is the depth range between 450 and 700 feet. This is where most of the deeper activity will be during the next two or three years. As we have seen, diving in this range will require the use of decompression schedules which are not well tested, and which will undoubtedly produce a high incidence of bends until they are empirically modified through experience. Therefore, jobs calling for diving in these depths must be supported with much more sophisticated equipment than has been required in the past.

For example, treating a case of bends that occurs on a 600-foot dive may require saturation decompression treatment. Obviously, the diving crew cannot spend days in a decompression chamber unequipped with adequate environmental controls and appropriate facilities for human comfort.

Summary

Since the majority of diving tasks from a drilling rig can be accomplished within 30 minutes total bottom time, it is possible, with good operational planning and experienced divers, to effectively use bounce diving techniques to about 600 feet. Tasks requiring longer bottom times can be accomplished with "back to back" diving techniques using a second deck decompression chamber and additional divers. Saturation can be used effectively from 300 feet, but would probably only be justified if unusually long bottom times were required. Such capability is more applicable to construction work rather than drilling. Below 600 feet, bottom times must be limited to avoid prohibitively long decompression times.

Control of compression rates will also use up available bottom time so that the effective working time is reduced to an impractically short period. The absolute limit for bounce diving is not easy to define, but for all practical purposes it is likely to be around 750 feet. The wise operator will provide for full saturation capability once he gets into water depths over 600 feet. The decision to utilize saturation techniques in this range will depend largely on the capability of the diving crew to accomplish the work within the time limits allowed, so they get through both the compression and decompression "windows" required for a bounce dive.

A ten minute job at 650 feet, if not completed in time, can force the crew to saturate in order to complete the work, thus tying up the decompression facilities for a six or seven day period. On the other hand, one might deliberately saturate the crew if the job is the first of an anticipated series of tasks (such as might occur if the marine riser is dropped and the BOP stack damaged). The divers would then be retained in saturation at a suitable storage depth, which would enable them to make rapid excursions to bottom pressures for extended working periods.

Once a depth of about 600 feet is reached, there is no present alternative to saturation. The use of sophisticated saturation diving equipment and highly qualified experienced personnel is mandatory.

DIVING EQUIPMENT AND PROCEDURES

Air Diving Equipment

Self-Contained Underwater Breathing Apparatus (SCUBA)

Scuba equipment utilizes high pressure cylinders to store compressed breathing air. The air is supplied to the diver by means of a demand regulator which delivers the air at a pressure equal to the surrounding water pressure. The diver's exhalation is exhausted into the water.

The advantages of scuba are horizontal mobility (because the system does not require any connections to the surface), portability as a result of its small size and weight, and a high degree of depth flexibility and buoyancy control. The disadvantages of the system are a depth limitation of about 60 feet (130 feet under ideal conditions), a limited gas supply, and an exertion limitation (caused by the mechanics of the demand breathing regulator). A serious operational disadvantage is caused by the fact that there is usually no communication between the scuba diver and topside personnel.

Scuba is seldom used in commercial operations because most underwater work does not require horizontal mobility, but almost always requires an unlimited breathing gas supply and good two-way communications for safety reasons.

Lightweight Equipment

Lightweight commercial diving equipment falls in two general categories. The first can be called facemask gear, which incorporates a full facemask supplied by an air hose from the surface. The mask also includes a communications system. With facemask equipment, the diver normally wears a standard neoprene wet suit.

The second category is helmet equipment, which is made up of a lightweight, surface-supplied fiberglass helmet similar in design to a hard hat. The helmet also incorporates two-way communications, as does all surface-supplied commercial diving equipment. Helmet systems may be used with neoprene wet suits (wet dress) or with dry suits. Both systems also consist of weight belts and boots or swim fins.

The advantage of lightweight gear is that it offers good mobility, minimum physical restrictions, an unlimited air supply, and good communications. Helmet gear also provides head protection and, when used with a dry suit, offers buoyancy control and good protection from cold water.

The necessary support equipment for lightweight gear includes an air compressor and volume tank, communications equipment and, if the job is deeper than about 100 feet, a decompression chamber. Air diving with this type of equipment is usually limited to 180 feet because of nitrogen narcosis.

Deep Sea (Hard Hat) Equipment

Deep sea or hard hat gear consists of a metal hard hat, a watertight suit (dry dress), a weight belt, and weighted boots. As with lightweight gear, an umbilical from the surface provides the air supply and communications. The advantages of deep sea gear are maximum physical protection, good working leverage (because the diver can make himself heavy underwater), and good temperature protection. The disadvantage of the equipment is its bulk and weight.

The depth restriction and surface support equipment are the same as for lightweight gear.

Mixed Gas (Helium/Oxygen) Diving Equipment

Mixed gas diving equipment is essentially the same as air diving gear, except that some mixed gas equipment incorporates a means of partial recirculation of the breathing gas in order to avoid exhausting expensive pre-mixed gas into the water. Lightweight mixed gas gear utilizes a back pack which contains the absorbent chemical that scrubs the CO_2 from the breathing medium. Deep sea mixed gas gear has the absorbent chemical in cannisters mounted in the helmet.

In addition to the usual air diving support equipment, mixed gas diving equipment requires the use of a manifold or regulator console which enables support personnel to switch the diver's breathing medium from air to helium/oxygen mixtures during descent and ascent. Also, mixed gas diving requires a supply of premixed high pressure gas and, in most situations, a voice unscrambler which will make the diver's speech intelligible when he is breathing helium. Mixed gas diving equipment is used at depths in excess of 180 feet in order to circumvent the nitrogen narcosis problem.

Submersible Chamber (Bell) Diving

Modern diving bells not only provide a refuge and source of breathing gas for the diver, but also serve as a means to return the divers to the surface under pressure, allowing them to take progressive decompression without remaining in the water. This is considerable improvement in diving safety during deep diving, and also represents a cost savings to the operator because rig down time is minimized. The divers can be brought directly back to the surface after their work is completed.

At this time, there are well over a hundred diving bells in operation around the world. Most of these are combined with deck decompression chambers (DDC's), to which they can be mated to permit the diving bell occupants to transfer to a larger, more comfortable pressure vessel for the long hours of decompression.

Most diving bells designed today are capable of withstanding external pressure without being pressurized inside, thus enabling the occupants to view the underwater work site through the port holes while remaining under atmospheric conditions in a relatively comfortable "shirt-sleeve" environment. Another advantage of this type of bell is that during pressurized lockout diving, the divers need not be put under pressure until the bell is properly positioned at the underwater work site. The diver's time under pressure can be brought to an absolute minimum and the decompression requirement will be as short as possible.

Job conditions and water depth generally indicate when a bell system should be used. As a rule of thumb, bell systems become economically feasible in depths greater than 250 feet and are higly desirable from a safety standpoint in depths in excess of 300 feet. Bell systems should probably be mandatory in depths greater than 350 feet. The cost of a unit is negligible when all other factors are considered.

Most bell systems are skid mounted and are made up of a DDC (Depth Chamber), bell, air compressor, hoisting winch and boom, and a mixed gas life support console. Most systems are designed so that the diving bell can be mated to or locked onto the DDC to allow the divers to transfer from the bell to the decompression chamber under pressure.

The diving bells are usually fitted with lights and an emergency supply of gas. Some bells are pressurized with gas stored in bottles on the bell and some are pressurized through an umbilical to the surface which also contains hard wire communications and power cables. Some bells have a side mating capability with DDC, while others are top mating.

Bell systems have made their largest impact in the area of oil exploration from floaters and semisubmersibles working in water depths in excess of 250 feet.

Saturation Diving Systems

A typical saturation system is made up of the following basic components:

Diving Bell

The bell is used to transport the divers from the topside habitat to the work site and return. It must be capable of mating with the living chambers.

Deck Chambers (DDC's)

The living chambers must provide space for sleeping, relaxing, work preparation, eating, and sanitary facilities. The systems are usually outfitted with transfer locks, T.V., radio, air conditioning and heating, and all life support functions.

Operational Motor/Control Module

Generally, this is a van containing communications and the equipment required to monitor and control the in-water dives.

Decompression Control/Monitor Module

This is a van which contains all of the controls and monitor equipment needed to operate the deck chamber complex and life support functions.

Support Equipment

Support equipment for saturation systems can vary, but most systems designed for full-scale deep water diving will include most of the following components:

Water supply system
Air compressor
Diver heating unit
Hydraulic power unit
Bell handling system (winch, tuggers)
Gas storage cylinders.

Saturation diving begins to make economic sense in water depths greater than 300 feet for projects requiring quite a few hours of working bottom time and in water depths greater than 200 feet when a relatively large number of working hours are required. For instance, several different jobs in 200 feet of water may prove to be more economical if done with the saturation technique.

Tools and Special Equipment

The real value of the diver is shown in his ability to perform useful work underwater. Most of the diver's tools are the same as topside tools, with certain modifications to render some of them watertight. Pneumatic tools, saws, drills, jackhammers, rivet busters, and impact wrenches are commonly used.

In addition to the equipment mentioned above, some of the more common diver tools are underwater cutting and welding gear, underwater lights, still and movie cameras, ramset tools, jetting equipment, air lift equipment, and most common hand tools.

Some special tools utilized by the diver are ultrasonic testing equipment, X-ray gear, water blast and sand blast equipment, magnetometers, pingers and sonar devices, and bottom coring devices.

DIVING SERVICES AND CAPABILITIES

Surveys and Inspections

This is one of the most common assignments and may range from a very simple bottom survey to ensure that there are no obstructions prior to setting a platform to very complex inspections to determine structural damage.

Observation bells and submarines are useful tools for survey and inspection work. Also, television, video tape, and photography provide permanent records. Ultrasonic equipment with strip recorders can be utilized to check wall thicknesses and critical weld areas with a record being maintained for future comparisons.

Surveys and inspections can be made quite accurately if the need exists. Limiting factors are poor visibility, depth, temperature, strong currents, and rough weather.

Maintenance and Repair

Routine maintenance of subsea installations is a common diving requirement. The diver can inspect for scour, leaks, coating damage and other discrepancies. Miscellaneous repairs to docks, piers, and other shore facilities will be performed by the diving team. These men are skilled craftsmen who can usually perform a multitude of duties other than diving.

Underwater repairs of every conceivable nature can be achieved. Damaged pipelines and platforms can be repaired by utilizing wet or dry welding techniques, or mechanical devices. The depth and scope of work will dictate the equipment, method, and cost of a particular repair. It would be safe to say that almost any repair can be made to existing installations in depths up to 600 feet. Dry welding is usually practical, though costly. Mechanical connections are proving to be reliable and will be used more and more.

The placement of concrete and repairs to concrete structures can be done with great success. Underwater cleaning and painting has reached the point where good results can usually be obtained.

New Construction

Pipelines

Diver support of pipeline construction operations is one of the more important roles played by today's diver. This role will expand in importance as operations move to deeper water depths. The divers are used to inspect the bottom for obstructions prior to laying the line. During the lay operation, the divers swim the stinger to ensure proper alignment of the pipe and to profile the stinger position. Quite often divers manipulate ballast valves to change the profile of the stinger.

The divers also check to ensure that coating damage is minimal and that the pipe is laying properly. During the burying phase, divers attach the jetting equipment and/or hand jet the line to the proper depth. Inspection divers are employed to check the line for proper installation.

Underwater connections and riser installations rely on the diver to a great extent. Familiarity with pipeline terminology and procedures is a prerequisite for this type of work. Pre-planning and pre-engineering of subsea connections and riser installations will go a long way towards reducing the costs and complexities.

Platforms

The basic task for a diver associated with platform installations is to check the bottom prior to setting the jacket. The role of the diver becomes more important when a platform is being installed over an existing wellhead. In this case, the diver assists in guiding the jacket into the proper position.

Harbors—Port Facilities

The major role of the diver in new port and harbor development is in the area of inspection. However, if rock is encountered, divers are often assigned the task of blasting away rock overburden or outcroppings.

Offshore Drilling Support

In all probability, the one area where the contribution of the diver is most significant is in support of offshore drilling operations utilizing subsea completion equipment. This is especially true for foreign exploration operations, which are often done in remote or isolated locations. To a large extent, the diver can be compared with a local fireman. You hope you never need him, but when you do, you really need him.

Basically, divers are utilized for the following tasks:

Check bottom and setting of base plate;

Stab guide wires for stack and for T.V. system;

Assist in stabbing conductors and marine risers;

Inspect BOP stack (when required) to check rams, kill, and choke lines and various hydraulic fitting and components. Locate and repair leaks;

Restab broken guide wires and clear debris from stack;

Manually operate components if hydraulic failures occur;

Assist in stack recovery.

A tremendous amount of rig down time can be saved by having a diver make an inspection or repair in lieu of having to pull and rerun the stack. Needless to say, rig divers must be thoroughly trained and familiar with whatever subsea drilling equipment is being used. They must understand the inner workings of a BOP stack and be aware of what can be done to correct malfunctioning parts.

Underwater Welding and Habitats

Hot taps, pipeline repairs and structural repairs can be performed with a high degree of effectiveness. A habitat which is large enough to accommodate two diver-welders and the structural member to be repaired is lowered to the work site, sealed around the pipeline, and then dewatered with compressed air or some other inert atmosphere.

The two most common welding techniques are known as TIG and MIG and require both training and skill to achieve code grade results. In addition, special equipment and gas shielding is required. A new method has been developed which requires much less training and special skills and compares closely with standard stick welding in the dry. This method utilizes manual metallic arc "stick" electrodes with special coating chemistry to stabilize the arc under hyperbaric atmospheres.

Ultrasonics, Television, Photography

From a technological standpoint, all of these services are well advanced and are far more valuable than many people realize. They extend the eyes of the diver to the topside engineer, technician or supervisor. Permanent records can be obtained and kept for later comparisons or documented evidence.

Underwater application of ultrasonics is a proven means of nondestructive testing. Wall thicknesses can be measured to determine the degree of corrosion. Critical welds can be checked for cracks and other deficiencies. Strip recorders are available for permanent recording of the data.

Underwater television can be one of the most useful and effective tools available to an oil company engineer. Provided the water is not too turbid, the engineer on deck can see what is being inspected and can often direct remedial action immediately. Video tape recordings allow studies to be made to determine what courses of action should be taken.

Although underwater photos do not provide an immediate picture of subsea conditions, they do provide a permanent record which can be studied in detail. Still and motion picture cameras have been developed which can produce amazing pictures in water with almost zero visibility.

Demolitions

The use of demolitions underwater has great applicability. Bulk and shaped charges can be designed for a multitude of jobs. Demolitions are used in salvage work, pipe trenching, harbor and port blasting and in removing conductors, wellheads and caissons below the mudline.

Submersible Work Vehicles

There are a number of small submarines on the market today. Depending upon the job, they can be very useful and effective. Most are expensive, but under certain circumstances they can provide the most economical means of accomplishing a task.

Beyond diver depths, subs can be extremely valuable. Subs are used for salvage work, inspections, surveys and scientific assignments.

REFERENCES

1. Dobbins, P., *Undersea Operations and Equipment*, School of Offshore Operations Presentation, Houston, Texas 1972.
2. Dugan, James, *Man Under the Sea*, The Macmillan Company, New York, 1965.
3. Earls, Thomas, *Deep Diving From Oil Rigs—What Are the Options?*, Offshore Scotland Conference Presentation, Aberdeen, Scotland, February 1973.
4. Hughes, D. M., *Deep Diving—Capabilities and Limitations*, Offshore Magazine, 1973.
5. Oceaneering International, Inc., *Rig Diving Data* (unpublished document), 1972.
6. Zindowski, Nicholas B., *Commercial Oil—Field Diving*, Cornell Maritime Press, Inc., Cambridge, Maryland, 1971.

PART V
Drill String Testing

18

Downhole
Well Testing

B. P. Nutter

Johnston-Schlumberger

The major objective to be achieved in testing an offshore well from a floating vessel is to evaluate the reservoir and provide the operator with valid information for making a good decision concerning future development.

The reservoir evaluation must be conducted with complete safety for personnel and equipment as well as protection of the environment. The principal considerations in achieving this objective are preplanning and preparation.

First, a detail test procedure should be established for obtaining good test data with complete safety as follows:

1. Personnel instruction and coordination planned before the test.
2. Selection of a tool system which requires a minimum amount of pipe manipulation for control and one which provides maximum reliability.
3. Data systems for pressure, fluid sampling, and metering.
4. A predetermined space-out procedure for proper landing of the pipe string.
5. Design of the test for safe surface pressure by proper use of a bottom hole choke, surface choke, water cushion, nitrogen cushion, etc.
6. Predetermination of the duration of the IF, ISI, FF and FSI periods.

7. Adequate means for handling crude oil and gas production.

Second, detailed equipment needs should be planned well in advance to insure availability, good condition, and adequate backup. For equipment requirements, the following should be considered:

1. Type of packer.
2. Bypass and equalizing means.
3. Placement of the pressure recording gages.
4. Screening of formation fluid.
5. Test control valve.
6. Reverse circulation means.
7. Safety valve systems.
8. Balance slip joints.
9. Jars and safety joint.
10. Subsea test tree.
11. Surface control head and manifold.
12. Surface metering, gaging, and separation.
13. Crude oil burner, flares, storage tanks, pumps, adequate air supply, etc.

Third, a contingency plan for a "fail-safe" operation should be established which would consider:

1. Excessive surface pressure.
2. Abnormal annulus conditions.
3. Weather, waves, tides, collision, anchors, etc.
4. Abnormal surface fluid problems, H_2S, leaks, etc.

BASIC TEST SYSTEM

The principal objective of the formation test is the same, whether it is conducted from a floating vessel or a stationary rig. The test, however, is performed with different techniques and tool systems for the two types of rigs.

When conducting testing operations on a stationary rig such as a jack-up, platform, or land type rig (Figure 5-1) there is no relative motion between the rig and the well in a vertical direction. On this type of rig, conventional testing systems which are

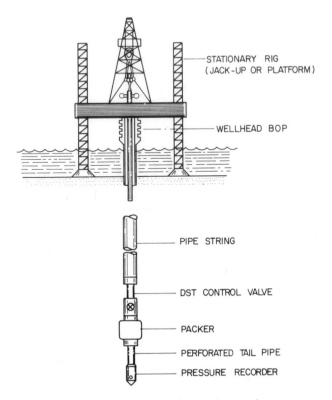

Fig. 5–1   *Testing system for stationary rig.*

operated by the up and down motion of the pipe string are used with a high degree of success.

On the other hand, when a well is being tested from a floating vessel there is a perpetual cyclic motion between the vessel and the well in a vertical direction due to wave and tide action. This motion can vary in degree depending on weather conditions. For this reason, it is evident that the conventional system which utilizes up and down pipe string motion for control is not as reliable on the floating vessel as on the stationary type rig.

A new testing technique and tool system (Figure 5-2) has been developed for conducting tests from a floating drilling rig. The

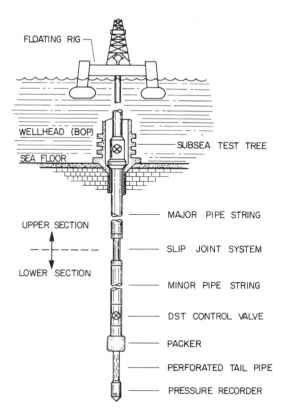

FLOATING RIG

WELLHEAD (BOP)

SEA FLOOR

SUBSEA TEST TREE

UPPER SECTION

LOWER SECTION

MAJOR PIPE STRING

SLIP JOINT SYSTEM

MINOR PIPE STRING

DST CONTROL VALVE

PACKER

PERFORATED TAIL PIPE

PRESSURE RECORDER

Fig. 5–2   *Basic test system for floating rig.*

new method basically differs from the conventional method in that:

1. The upper end of the pipe string is supported by the ocean floor wellhead rather than the rig,
2. The principal control valve in the testing string is operated by annular pump pressure applied from the surface rather than manipulation of the pipe string.

The new testing string is divided into an upper section and a lower section. The lower section consists of pressure recording gages, perforated or slotted tail pipe, a packer, a formation flow control valve, and the minor pipe string. During the test, this lower section is supported by the ocean floor wellhead.

A pressure-volume balanced slip joint system divides the upper and lower sections of the testing string and serves as a combination expansion and contraction joint during the test. Proper space-out when setting the packer and landing in the wellhead places the slip joint system in a mean position. Any shortening or lengthening of these two sections of the string as a result of temperature and/or pressure changes during the test is compensated for by the slip joint system. This maintains the weight of the minor string on the packer.

The principal downhole flow control valve in the new string is operated by annular pump pressure applied from the surface. Once the packer is set and the upper section is landed in the ocean floor wellhead, no further movement of the pipe string is required to conduct the test.

PCT Testing String

For a better understanding of the new testing system, Figure 5-3 illustrates the relative location of each component in the string. Starting at the lower end, the string is composed of:

1. Pressure recording gages.
2. Perforated or slotted tail pipe.
3. Packer.
4. Safety joint.
5. Jars.
6. Upper recorder housing.
7. Hydrostatic reference tool.
   a. Pressure reference valve.
   b. Backup flow control valve.
   c. Integral bypass valve.
8. Pressure controlled tester.
9. At least 90 feet of drill collars for sand which may settle from produced fluid.
10. Cushion displacement valve for spotting cushion.
11. Rotary type reverse circulation valve.
12. Pressure (internal) actuated reverse circulation valve.
13. Minor string (collar weight required to set packer and operate MFE tool).
14. At least one slip joint.

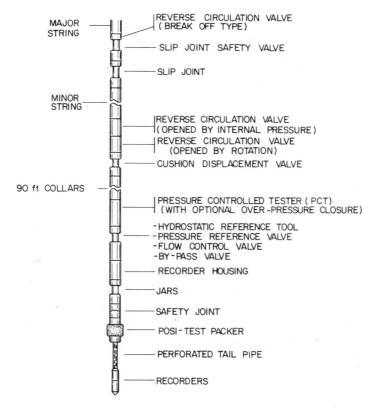

Fig. 5–3 *PCT tool string.*

15. Slip joint safety valve.
16. At least 2,000 lbs. of collars between slip joint and slip joint safety valve.
17. Break-off type reverse circulation valve.
18. Major pipe string.

Wellhead and Surface Pressure Control

Figure 5-4 illustrates a typical ocean floor wellhead (BOP stack), a subsea test tree, and surface pressure control equipment used when conducting testing operations from a floating drilling vessel.

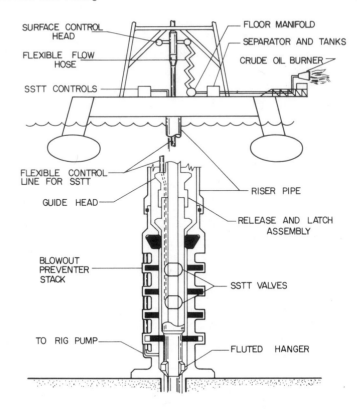

Fig. 5–4   *Wellhead and surface controls.*

TYPICAL TEST PROCEDURE

Space-Out and Landing Procedure with Posi-Test Packer

During the test, the length of both the major and minor pipe strings are subject to change. These lengths can either increase or decrease, depending on the changes in temperature and pressure. A minimum of two (2) slip joints are run between the two strings to allow for this expansion or contraction. One slip joint, provided with a safety valve, is installed on the lower end of the major pipe string. The other slip joint is installed on the upper end of the minor string. One (1) drill collar is run between the two slip joints to assure proper sequence of operation.

The space-out operation should function so that when the packer is set, the fluid displacement valve is closed, the hydrostatic reference tool is open, and the major string is landed in the ocean floor wellhead, the upper slip joint will be fully extended and the lower slip joint would be fully contracted.

The following description is an analysis of the complete testing string which takes into consideration all of the tools that function as a slip joint. An ideal space-out and test is also described.

First, determine the total travel of all slip joints which will be contracted or closed during the test, starting at the bottom and working upward, as follows:

Hydrostatic reference valve	0.5 feet
Fluid displacement valve	0.5 feet
Lower slip joint (at top of minor string)	5.0 feet
Buckling and compression of minor string	1.0 feet
	7.0 feet

This total of 7.0 feet is the tool space-out length.

Next, locate the landing sub or fluted hanger in the string to land with the packer at setting depth plus the tool space-out length (7.0 feet).

Install the control head.

Note the weight indicator reading while packing up.

Slack off and land the fluted hanger sub in the BOP without setting the packer. Make a chalk mark on the pipe with reference to the riser. Pick up and make a second chalk mark, 7 feet below the first mark to indicate the setting position of the tools.

Check wave action (determine elevator movement relative to pipe), with pipe landed in the wellhead.

At the bottom of a wave, latch the elevators and pick up, torquing to the right for 7 feet (to the lower chalk mark), then slack off allowing string to land in the wellhead. Immediately determine the weight supported by the wellhead with a slight pick up and slack off. This weight should be less than the initial weight because of the elimination of the weight of the minor string and tools which are now supported by the packer.

Allow at least 10 to 15 minutes for the hydrostatic reference tool to operate (hydraulic time delay) since no surface indication is produced.

Operating the PCT Tool

The PCT is opened for the initial flow by closing the rams and applying from 1,000 to 1,500 psi pump pressure to the well annulus. A very weak blow will be observed during this pressurization operation due to the squeeze or collapsing effect of the pump pressure on the pipe string. This should not be confused as an indication of the PCT opening.

The PCT is closed for the initial shut-in by bleeding the pump pressure on the annulus back to zero. The PCT is reopened for the final flow by pressuring back up to 1,000 to 1,500 psi on the well annulus. During this period, as the surface flow pressure increases, the annulus pressure will also increase due to the expansion of the pipe string and increase in temperature. The annulus pressure should be maintained between 1,000 and 1,500 psi. After the water cushion is unloaded and the flow pressure becomes stable, the pressure in the annulus will also stabilize.

If, during the flow period, the annulus pressure continues to increase after the flowing pressure has stabilized, a leak in the pipe string is indicated. Continued bleeding off of this pressure will eventually result in the well flowing through the annulus. If this occurs and a pipe leak or failure is indicated keep the annulus closed in, thus allowing the pressure to build up. When the annulus pressure reaches a level of 2,500 psi, the PCT will close as the over-pressure system is activated. This automatically shuts the well off at the bottom.

If the surface pressure on the annulus does not reach the 2,500 psi level the pumps should then be applied. The closure of the PCT on bottom will be indicated by a gradual reduction in both the tubing and annulus pressure at the surface.

Two methods can be employed for clgsing the PCT for the final shut-in: the pump pressure on the annulus can be bled off, or the pressure on the annulus can be increased with the pumps to approximately 2,500 psi. The latter is preferred since the tool becomes permanently closed and will not accidentally reopen if excessive pump pressure is required during reverse circulation.

Reverse Circulation

After the test and prior to pulling the tools from the hole, all recovered formation fluids must be removed from the string and replaced with mud for safety and prevention of pollution. This can be done either during or after the final shut-in period. It is preferable to reverse circulate during the final shut-in period since valuable rig time is saved and sand (produced into the pipe string) will be kept in motion and not allowed to settle out during this period.

The reverse circulating valve should be opened immediately (within 5 minutes) after the tool is closed for the final shut-in. If opening cannot be achieved within this 5 minute period, then reverse circulation should be delayed until the shut-in period is completed. It should be emphasized that the hydraulic and/or mechanical impact produced when opening the reverse circulation valve often causes an anomaly or disturbance in the pressure buildup curve.

In many cases, if the disturbance occurs during the steady state portion, the buildup curve cannot be used for reservoir analysis involving the Horner Plot. Two valves can be opened for reverse circulation during the final shut-in: the pump-out type which is opened by pressuring up on the interior of the test string, and the break-off type which is opened by dropping an impact bar.

When reverse circulating is performed after the final shut-in, two other valves in addition to the preceding valves can be used: 1) if the fluid cushion displacement valve has been run without the lock it can be opened by simply picking up the test string, and 2) the rotary valve which is opened by rotating the test string 10 revolutions to the right. The rotary method should be used only as a last resort for the following reasons:

    a. Rotation may damage the flexible line which controls the SSTT.

    b. The string must be picked up out of the wellhead and supported by the rig to rotate.

    c. The rams must be open to rotate.

d. It is not safe practice to rotate with high surface pressure on the test string.

If for some reason none of the preceding reverse circulating valves can be opened, then a lubricator is rigged up on the control head and a tubing gun is run for perforating the test string. After retrieving the gun, the fluid is reversed from the string.

Pulling Loose

As the pipe string is lifted in the pulling loose procedure, the bypass valve in the HRT opens and equalizes the pressure across the packer. This assists in getting the Posi-Test Packer loose or in pulling the seal nipple or stinger out of the permanent packer.

SUMMARY

To summarize, a new testing system has been developed for conducting formation testing operations on offshore wells as shown in Figure 5-5. The principal flow control valves in this new string are operated by pump pressure applied from the surface. For this reason, it is ideally suited for conducting formation testing operations from a floating type vessel or in a highly deviated hole where controlled pipe string manipulation is very difficult or impossible.

For optimum protection of personnel, equipment, and the environment the new testing system is provided with a family of fail-safe control valves for shutting off the flow of formation fluid to the surface. The valves can be closed by various methods.

1. The pressure controlled tester closes when annular pump pressure is bled off.
2. The pressure controlled tester automatically closes if the annulus is over-pressured.
3. The MFE control valve closes when the pipe string is lifted.
4. The slip joint safety valve closes if the major string parts and drops.
5. The subsea test tree valves are closed by bleeding off special hydraulic pressure at the surface.

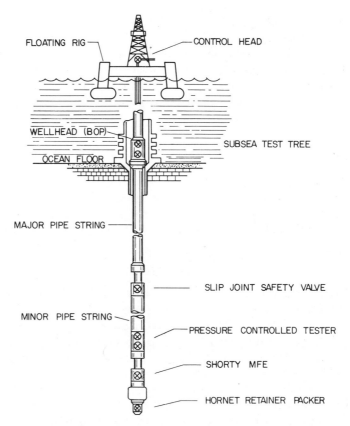

Fig. 5–5 *Formation flow control valves for PCT test system.*

6. The surface control head and floor manifold can be closed by manual and remote means.
7. If a cement retainer is used, the valve in the retainer closes when the pipe string is lifted.

Nomenclature

IF:	Initial Flow	SSTT:	Subsea Test Tree
ISI:	Initial Shut-in	HRT:	Hydrostatic Reference
FF:	Final Flow		Tool
FSI:	Final Shut-in	MFE:	Multiflow Evaluator
PCT:	Pressure Controlled Tester	DST:	Drill Stem Test

19

Production Testing and Burners

J. E. Weatherly, Jr.
Weatherly Engineering

WELL EVALUATION

In evaluating a well's productive capacity, the type of test and testing parameters that are usually involved include:

Types of tests

1. *Oil well—regular*
2. *Gas well—regular, AOF (absolute open flow), or flow rates*
3. Extensive
   a. Destructive
   b. Reservoir limit
   c. Remedial evaluation

Parameters

1. *Surface pressure anticipated*
2. *Volumes anticipated*
3. *Production characteristics*

Recommendations for surface equipment

1. *Gas condensate requires heater and separator*
2. *Oil wells usually require separator only*

3. *Auxiliary equipment*
   a. Tanks
   b. Transfer pressure flow lines
   c. High pressure flow lines
   d. Recorders
   e. Dead weight gauges
   f. Chloride kits
   g. Hydrometers
   h. Shake-out machines
   i. Hydrogen sulfide apparatus and dectors
   j. Burners
   k. Sample bottles for PVT, gas and liquid analysis

Recommended subsurface equipment

1. *Bottom hole pressure gauges*
2. *Bottom hole samples*
3. *Hangers*
4. *Wireline units*

Other factors which can govern the choice of testing equipment are the size and type of rig or platform on which the equipment will be located. Equipment such as burners is rigged differently on jack-ups as opposed to semisubmersibles or drillships, and domestic equipment is usually different from large capacity international testing apparatuses.

Normally, for Texas and Louisiana Gulf Coast testing, a three to four thousand barrel per day fluid capacity and ten to twenty million cubic foot gas capacity, three phase 1,000 to 1,500 psig working pressure separator will suffice. Sometimes a second stage is incorporated on the unit which operates at 125 psig working pressure. Measurement of fluid is performed with either positive displacement or turbine meters on both stages and gas is measured with orifice meters to AGA standards.

In conjunction with the separator, when gas condensate production is anticipated, a one million BTU, 10,000 psig working pressure heater is used to prevent freezing in the separator. Depending on conditions and length of testing required, the operator may have the testing company furnish a variety of auxiliary equipment such as tanks, transfer pumps and burners.

Since this equipment by its nature is quite heavy, capacity

limitations are governed by crane capacity and room on the rig. Additional capacity requirements can be met by adding units and splitting the stream.

In most cases this is a variable load on the rig and is loaded with proper lead time and unloaded after completion of the test program.

However, on most international rigs this is not the case. The test units that will be in remote sections of the world usually become a static load on the rig and are installed as a miniature tank battery and separation facility. International equipment is tailored to the specifications outlined by the operator and designed to conform to the space allocated by the drilling contractor. The designs incorporate climate and international regulations. For instance, North Sea equipment is different from equatorial designs.

Space on the rig is always a major problem. Double decking over other areas on the rig has been used and piggy-backing of equipment has also been adopted to conserve space. For a recent North Sea design, approximately 80 tons of static load was added for test equipment. High concentration of loads have to be considered. For example, an eight foot diameter 100 barrel capacity test tank full of salt water will weigh 47,000 pounds.

Personnel furnished by testing companies varies; however, the crew should consist of a senior tester and helper on a Gulf Coast test without a burner. Where a burner or burners are involved, both domestically or internationally, the crew should consist of a senior tester, junior tester and helper-burner operator combination. These people should be competent in both surface and subsurface testing which includes operating and gathering pertinent data from surface and subsurface equipment.

A summary of the data is presented to the company involved and final calculations and reports are transmitted in usable form.

Safety Precautions

Safety cannot be emphasized strongly enough. Test equipment should be hydrostatically tested by rig pumps on a periodic basis and particularly prior to testing.

Since the equipment is subjected to the most devastating conditions on the original completion test, it is designed in such a manner to avoid unnecessary shutdown due to equipment failure. Erosion and corrosion of steel due to sand and mud production along with corrosive well conditions are the greatest causes of failure. Therefore designs are simple, and sophisticated controls are not incorporated due to the rugged production conditions. Vessel interiors are simple, aimed primarily at decelerating the well stream to give adequate separation. Mist extractors, screens, baffles, and iron sponges are usually not incorporated. After one to two years service all connections should be inspected and vessels either magna-fluxed or x-rayed.

Certain indicators for hydrogen sulfide units normally carry air packs for personnel to use when exposed directly. Indicators will detect the danger level of hydrogen sulfide should small leaks contaminate closed-in areas.

Safety valves can be installed on the floor at the flow line entrance with a panic button to actuate a valve close to the test unit. On larger semisubmersibles and drillships the test equipment is usually remote from the moon pool. Therefore an intercom system is most advantageous in conjunction with the safety valve for good communications throughout the rig.

In some cases a safety valve is installed at the heater ahead of the adjustable chokes which is actuated if the surge tanks become over-loaded. This would automatically shut in the well at the testing station. Other sensors could be incorporated, but due to the nature of the testing a minimum amount of automatic valves should be used. A point which should be noted is that during the testing the equipment is fully attended with competent testers that can detect irregular or erratic conditions and implement corrective measures prior to accident or failure.

A meeting with the rig personnel involved in the testing is mandatory prior to test. Coordinated efforts can reduce the possibility of an accident.

Competent experienced personnel is the answer to a good and safe operation. There is no substitute for good people.

RULES AND REGULATIONS

Testing procedures must adhere to all rules and regulations prescribed by regulatory bodies. Examples are described below:

Railroad Commission of Texas

1. *Rules pertaining to type of test required to obtain allowables:*
   a. Gas wells
   b. Oil wells
   c. Gas-condensate wells
2. *Safety Rules:*
   a. Open flames
   b. Flame arrestors
3. *Department of Wildlife, Parks, and Recreation*

Louisiana Conservation Department

1. *Testing procedures to obtain allowable*
2. *Stream control commission*

U.S.G.S.

1. *Representative required on initial testing in federal waters*
2. *Pollution control (drip pans, etc.)*
3. *Types of testing required to evaluate leases shut in gas well regulation*

U.S. Coast Guard—Various Codes for Equipment Specifications

1. *ASTM*
2. *ASME*
3. *NEMA*
4. *API*

Most of these requirements are stipulated while testing in federal waters and are prescribed before obtaining various types of insurance.

International Laws and Treaties—
Protection of Environment from Pollution

To protect our environment from pollution, the testing companies have been called upon to develop and incorporate various anti-pollution devices. Simply, these are spillage containment devices and burners to be used where extensive high volume testing is going to be performed. NATO has signed a treaty relating that by 1974 spillage, willful pollution, and bilge pumping would not be tolerated anywhere in the Free World.

BURNERS

Pollution Control International, or PCI, Inc., has developed a super oil burner. Along with Baker John Zinc burners designed in England, and the Flopetrol (subsidiary of Schlumberger) giant burner developed in France, it is designed to test wells without retentive storage and eliminate pollution.

Basically, burners involve atomization and air mixtures to obtain, hopefully, 100% combustion. Designs and specifications vary for smokeless burns, and for volumes anticipated. The burners currently used are 5000 barrels/day and 20 million cubic feet/day gas capacity and 10,000 barrels/day and 35 million cubic feet/day gas capacity. The difference lies in designing piping to handle the fluids and gas, and of course the water and air requirements.

Burning through super oil burners is relatively new to the oil industry and much is still to be learned. Various designs are on the drawing board in which rig-up time, storage on rigs, and multiple use of booms are being considered. The devices are expensive and are high maintenance items just due to the heat and abrasive material that flows through the burners. Part of the problems have been cured by using surge or test tanks and pumping with transfer pumps at steady rates from the tanks into the burner.

Burning off-the-side of rigs is a problem in controlling the radiated heat back to the rig. This is being controlled somewhat by water sprays. Designs of booms and guy wire arrangements will vary for jack-ups as opposed to floaters or semisubmersibles. Consideration in mechanical design must be given to "G" forces on unstable rigs; otherwise impact and fatigue could snap the boom. Wind forces should also be considered in structure design. Changes in wind direction can have a very adverse effect while testing, therefore most operators rather than interrupt a test procedure are requiring two complete burners on opposite sides of the rig and manifold for immediate switchover.

Pilot systems vary with burners. Our original concept was to drift away from the volatiles such as butane, propane, etc., mainly because they are not readily available in some sections of the world and because of the restrictions in handling. Therefore, we developed an electrically ignited diesel oil pilot system and it appears to be working satisfactorily. Other devices are available that can switch the natural gas being produced to the pilot system to conserve diesel where possible.

PART VI

Certification and Insurance

20

Classification and Certification of Offshore Drilling Units

Peter M. Lovie
Engineering Technology Analysts, Inc.

Offshore drilling rigs are often purchased on the basis of the unit's "classification." Insurance rates are dependent as well on the classification of the unit. The governments of Norway and Britain also now require offshore units to be certified for fitness before they can drill in their waters. What procedures are involved in classification and certification of a drilling unit? Secondly, what are the requirements for obtaining the appropriate classification or certification?

This section will serve as an introduction to the regulatory bodies who set the standards for offshore mobile drilling units and as a guide to their functions. It does not seek to describe or compare the technical requirements established in their "Rules." Further information may be obtained from the agencies themselves or from their publications (see Appendix).

The need for the certification of performance for offshore drilling units arises from concern for three factors:

1. Safety, reliability: Drilling contractors and operators want to protect their profits (and their reputations).
2. Avoidance of financial losses: Insurance agencies want to minimize the risks involved in offshore drilling operations.

3. Prevention of loss of life (particularly in the hazardous conditions of the North Sea): Government agencies want to ensure the safety of personnel on the rigs.

The functions of the various classification societies and governmental regulatory bodies often overlap. The major difference between the two is that while the regulatory bodies are legally empowered to enforce their rules, the classification societies are independent organizations whose power is strictly commercial. But the governmental agencies are turning more and more to the classification societies as consultants or as their designated representatives. The rig owner and operator must necessarily be familiar with the workings of both organizations.

CLASSIFICATION SOCIETIES

The offshore drilling industry has adopted the practices of the shipping industry in accepting the established authority of the classification societies as certifying organizations. The worldwide classification societies are American Bureau of Shipping (U.S.A.), Bureau Veritas (France), Det Norske Veritas (Norway), Germanischer Lloyd (Germany), Japanese Marine Corporation (Nippon Kaiji Kyokai) (Japan), Lloyd's Register of Shipping (U.K.), and Registro Italiano (Italy). The classification societies primarily involved in the survey of offshore mobile drilling units are the American Bureau of Shipping, Det Norske Veritas, and Lloyd's Register of Shipping.

The original need for a classification society for marine vessels arose around 1760. A group of marine underwriters meeting in Lloyd's Coffee House in London, England decided that they needed a means of uniformly comparing the capacities and conditions of ships for insurance purposes. The first register was printed in 1764, listing the names of ships and the dates when they were built. The condition of the hull was listed as A, E, I, O, or U (with A representing the highest possible rating and U the lowest). Equipment was classed as A, M, or B (good, middling, or bad).

As time passed, more and more detail was incorporated into the classification of the vessels. The hull and equipment classification systems were later changed to have "A1" representing the best possible rating—hence the famous "A1 at Lloyd's" classification. The standards for inspection and engineering analysis involved in the classification of vessels have changed and increased tremendously over the years. A registry book is still printed annually, containing the names and data for those ships classed by the society. Such information is also gathered on other ships of 100 gross tons or more which are not classed by the society.

Lloyd's Register of Shipping is organized as a nonprofit, independent corporation. It has no governmental connections. The organization of other classification societies, such as the American Bureau of Shipping or Det Norske Veritas, is patterned after that of Lloyd's Register. Each will perform survey and classification services for marine vessels and offshore mobile drilling units of other countries as well as those of their own. Each maintains a worldwide network to provide these services. The American Bureau of Shipping has survey stations in 221 cities in the world, Det Norske Veritas has 234, and Lloyd's has 252.

Begun in 1870 as the American Shipmasters' Association, ABS had classed 122 offshore mobile units of various types as of August 1974 with 79 units being built or committed for construction to ABS standards. Det Norske Veritas, which was organized in 1864 for the classification of ships, had classed seven operating offshore mobile units from 1970 and had 38 units being constructed to its requirements as of November 1974. Lloyd's Register had classed 17 operating units with 10 units being constructed to class as of November 1974.

The rules of the classification societies reflect satisfactory experience in service gained from comparing known types of vessels for certain parametric ranges. The various classification agencies differ in the depth of detail their regulations employ and, in some cases, their relative conservatism. However, they are alike in many respects. Their rules cover such items as:

1. Submission of detailed plans including midsection cross sections; construction details; framing plans; inner bottom plating; deck plans; watertight bulkheads; machinery,

boiler, and engine foundations; steering gear; ventilation systems; piping and electrical systems; anchor systems, etc.;

2. Load line curves;
3. Ship hull proportions;
4. Structural member sizes;
5. Other items concerning the structural integrity and seaworthiness of the unit.

There are separate rules for the classification of vessels for the open seas, for service on rivers and intercoastal waterways, for offshore barges, etc. The relevant rules for mobile drilling units are cited in the Appendix.

The classification of a unit is frequently used as a performance standard, specified in the construction contract, that must be met by the builder prior to acceptance of the unit by a drilling contractor. Units built under the supervision of one of the societies receive the distinguishing mark of the Maltese cross inserted before their classification. Thus, a first class jack-up built under the survey of Lloyd's Register will be classed ✠ OU100A1; under ABS it will be designated ✠ A1 Self-elevating Drilling Unit; and under DNV it will be ✠ 1A1 Self-elevating Platform. Of course, a previously constructed unit can receive classification from the societies if it meets the proper requirements; this classification will not bear the mark of the Maltese cross, however.

Classification is also a standard preferred by insurance underwriters. The classification indicates the standards to which the unit was built, thereby offering an evaluation of the risk of damage. An unclassed unit will require additional surveys, and insurance rates may therefore be more expensive than for a classed unit.

The classification societies perform other functions in addition to the classification of ships and offshore drilling units and the publishing of rules for the classification and construction of hulls and machinery. They also:

1. Supervise the construction of vessels/offshore drilling units, involving the analysis of designs, the witnessing of testing of materials, and the verification of building standards;

2. Survey the completed vessel/offshore drilling unit periodically throughout its economic life for the maintenance of
   classification;
3. Carry out load line surveys, safety equipment surveys, and
   issue certificates of character and tonnage and construction certificates under the authority of various governments and conventions;
4. Publish an annual register of the essential hull and
   machinery details, as well as the performance record, of
   classed vessels and units;
5. Perform continuous research on environmental conditions
   and design criteria;
6. In many cases, publish results of their efforts in periodical
   literature such as the *ABS Surveyor*, the DNV *Veritas*, and
   the *Lloyd's Register)*;
7. Maintain membership in the International Association of
   Classification Societies.

GOVERNMENTAL (REGULATORY) AGENCIES

Most nations have an arm of the government (such as the U.S.
Coast Guard, the U.K. Department of Energy, and the Norwegian
Maritime Directorate) which is responsible for the safety of ships
and offshore drilling units operating in their waters. While the
classification societies are concerned mainly with the structural
integrity and seaworthiness of the vessel, the governmental
bodies are concerned primarily with the safety of the people on
board and, increasingly, with environmental concerns.

The U.S. Coast Guard, for example, requires that ships and
offshore drilling units operating in U.S. waters meet its standards for safety equipment (such as fire fighting and lifesaving
equipment) and for minimum living quarters and sanitary conditions, and that the unit is operated by a certified crew. The
U.S.C.G. is also concerned with environmental factors such as
oil spills and shipping accidents. The British Department of
Energy (formerly the Department of Trade and Industry) has
responsibilities similar to the U.S.C.G. but will also enforce
overall safety and performance standards.

Until recently, there had been no *legal* requirements on the overall performance and safety of offshore mobile drilling units, although commercial requirements had been set through the classification societies. This has now changed. North Sea experience has shown that it is often impossible to evacuate a rig in the face of severe storms. Such storms can develop with as little warning as 2 to 3 hours instead of the 2 to 3 days' warning usually given by Gulf Coast hurricanes. Many of the North Sea locations are too remote for helicopters to easily operate in. Consequently, the drilling unit has to be adequate to ride out the worst expected storms without loss of life.

Much of this concern for safety stems from the loss of the "Sea Gem" in U.K. waters in 1967, a time when North Sea operations were experimental to a large degree. Recent observations of the real severity of North Sea operations have also spurred governmental interest in the safety of offshore units. Besides creating increased hazard to the offshore personnel, these rougher conditions compound the problem of environmental pollution and make the effects of accidents in any phase of offshore drilling operations much worse than in milder climates.

It must not be forgotten that although much of its offshore drilling activity takes place in remote, inaccessible areas, the North Sea is at the center of a highly industrialized and populated environment already heavily exposed to pollution. There is a growing sensitivity on the part of the general public as well as the governments involved in North Sea drilling to maintaining high environmental standards.

Unlike the "Rules" of the classification societies, the governmental regulations also cover production platforms. Additionally, they specify the environmental conditions (e.g., wave height, wind speed, current velocity) to be encountered in a given location. Figures 6-1 to 6-5 show examples of such information reproduced directly from the Department of Energy's publication "Guidance on the Design and Construction of Offshore Installations 1974." Thus, offshore mobile units may be approved for operation in certain locations only under specified conditions (summer only, year round, etc.). As a consequence of this, it may prove advantageous to the drilling

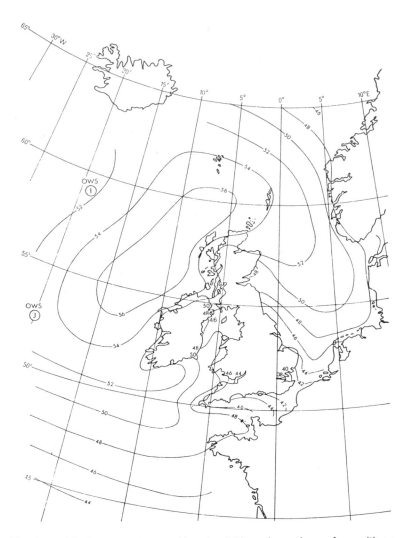

Fig. 6–1   *Maximum gust speed in m/s at 10 m above the surface with an
average recurrence period of fifty years. (Revised May 1973)*

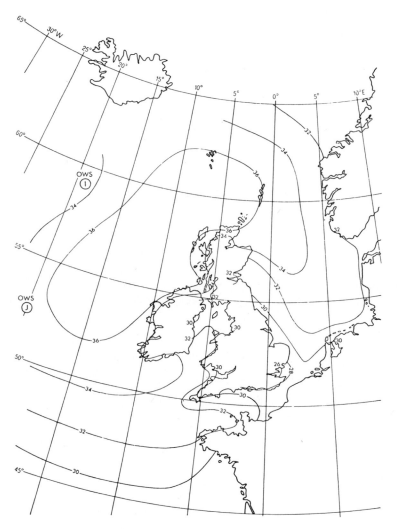

Fig. 6–2   *Hourly mean wind speed in m/s at 10 m above the surface with an average recurrence period of fifty years. (Revised May 1973)*

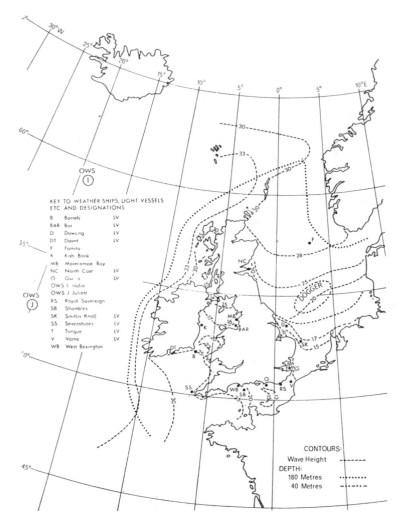

KEY TO WEATHER SHIPS, LIGHT VESSELS
ETC AND DESIGNATIONS

B	Barrels	LV
BAR	Bar	LV
D	Dowsing	LV
DT	Downt	LV
F	Famita	
K	Kish Bank	
MB	Morecambe Bay	
NC	North Carr	LV
O	Ow s	LV
OWS I	India	
OWS J	Juliett	
RS	Royal Sovereign	
SB	Shambles	
SK	Smitlis Knoll	LV
SS	Sevenstones	LV
T	Tongue	LV
V	Varne	LV
WB	West Bexington	

CONTOURS:
Wave Height ------
DEPTH:
180 Metres ··········
40 Metres —··—··—

Fig. 6–3   50-year design wave heights for a fully developed storm lasting
12 hours. Based on instrumental measurements and forecasts from wind
data. (Revised July, 1972) Wave heights in meters.

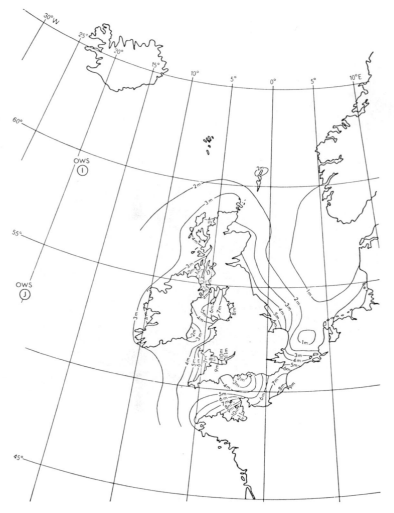

Fig. 6–4 *Lines of equal mean spring tide range (shown at 1 m intervals).*
(Revised December 1972)

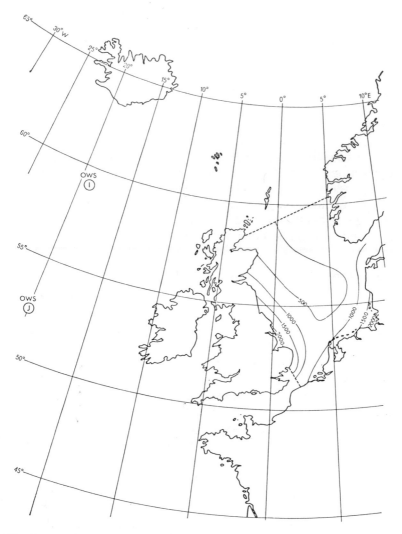

Fig. 6–5 *50-year storm surge (contours at 500 m intervals). (Revised December 1972). This information applies only to the areas between the boundaries shown.*

contractor to obtain certification for several different sets of criteria for an individual rig.

The determination of what storm criteria to use in the North Sea has been a matter of considerable controversy. Different oceanographers have predicted radically different criteria, although each may have been extremely competent and used well accepted analytical techniques. Intensive studies are now underway to establish accurate meteorological and oceanographic data. It is hoped that as more data are collected, criteria will be agreed upon that will not run the risk of penalizing the contractors and operators.

The measured storm criteria have increased so much in recent years in the North Sea that many operators believe that insurance underwriters and governmental authorities are now inclined towards using criteria that err on the severe side until strong justification is found for reducing the severity of criteria for wind, wave, and current effects.

Certificate of Fitness

The governments of the countries adjacent to the North Sea area which have drilling activity in their waters have made it a legal requirement that certain standards will be met before drilling operations can begin. The first government to require drillers to meet its safety and performance standards was the British Government through its "Mineral Workings (Offshore Installations) Act 1971." This act authorized the establishment of regulations for the certification of drilling units.

The "Offshore Installations (Construction and Survey) Regulations 1974" establish the relevant technical requirements for offshore drilling units (mobile and fixed) and specify that each offshore installation operating in U.K. waters must possess a certificate issued by a certifying authority stating that it is fit for use in those waters. Such certificates may be granted at any time prior to the unit's entry into U.K. waters but must be held at the time of entry. (A 16-month period is allowed for obtaining the certificate by units already operating in U.K. waters.)

The regulatory agency for certification of fitness in U.K. waters is the Department of Energy, which has authorized five classification societies to act as the certifying authorities: the American Bureau of Shipping, Bureau Veritas, Det Norske Veritas, Germanischer Lloyd, and Lloyd's Register. Similarly, the Norwegian Maritime Directorate now issues approval for both the design and operations of all units operating in Norwegian waters, with Det Norske Veritas serving as principal consultant. This does not mean that classification by the classification society involved is necessary for certification. Nor does certification by any of the societies acting as the certifying authority imply the granting of classification.

Other governments in Northern Europe are expected to follow the lead of the U.K. and Norway in establishing certificate of fitness requirements, once these rules have been in effect for some time. There is hope among the operators that a unified "certificate of fitness" authority or level of performance will be established for the entire North Sea area, although this will probably take place far in the future.

Other worldwide drilling areas do not have governmental regulation of overall safety and performance of drilling units, although for U.S. waters the U.S. Coast Guard and OSHA (Occupational Safety and Health Administration) requirements are legally enforceable.

SOLAS 1960 Applications

The administration of the government under which a vessel or offshore drilling unit is registered has the responsibility for the application of the requirements of the International Convention for the Safety of Life at Sea 1960 (SOLAS). In many cases, however, the governmental agencies of countries signatory to the convention have authorized the classification societies to survey a new or existing unit for compliance with SOLAS requirements and to certify the unit as to its compliance with the provisions.

It is therefore important to establish the unit's port of registry early in order to determine the appropriate SOLAS applications. Again, it is important to note that compliance with SOLAS is a requirement for operation and not for classification, and that classification does not necessarily imply that SOLAS requirements have been met.

REQUIREMENTS FOR CLASSIFICATION
AND CERTIFICATION

The first publication of rules pertaining to offshore mobile drilling units was the American Bureau of Shipping's "Rules for Building and Classing Offshore Mobile Drilling Units" in 1968. Since that time, the standards prescribed have been subjected to the test of actual application to a wide range of designs and a revised edition was published in 1973. Lloyd's Register published its "Rules for the Construction and Classification of Mobile Offshore Units" in 1972. In 1973, Det Norske Veritas published the "Rules for Construction and Classification of Mobile Offshore Units" developed from the preliminary "Principles for Classification of Offshore Drilling Platforms," published in 1970.

The societies are keeping up with the technological developments in offshore oil and gas exploration and production. The American Bureau of Shipping's "Guide for the Classification of Manned Submersibles" establishes requirements for the service capsules and work chambers utilized in subsea completion systems. Det Norske Veritas was the first classification society to propose tentative rules for the design, construction, and inspection of permanent offshore structures, including steel and concrete production platforms and storage tanks.

The classification and certification requirements for the detailed design and construction of an offshore mobile drilling unit are spelled out in detail in the various publications (see Appendix) and will not be described here. There are some important factors apart from rig design or construction which affect rig moves and operations, however, which should be mentioned.

Stability during Tow

This is expressed graphically in Figure 6-6. The unit must have an adequate inherent tendency to right itself in the water

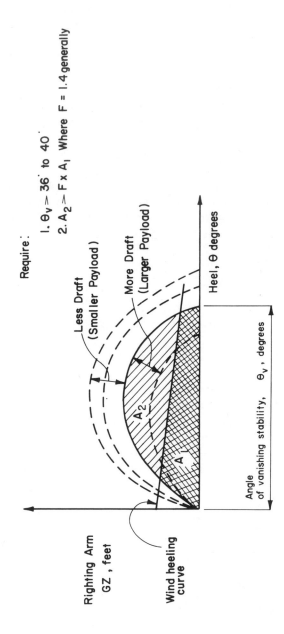

Fig. 6-6 *Stability requirements.*

whenever it is heeled over by an external overturning affect. This is expressed in terms of a righting arm, GZ. Counteracting this internal self-righting effect are the external overturning effects acting on the unit, which typically are dominated by the wind heeling effect.

The two "rules of thumb" used to ensure adequate stability are shown in Figure 6-6, as well as the effect on stability of adding weight to a unit: as weight is added (and the draft increases), the stability of the unit decreases. Although it might appear that the deeper a vessel sits in the water the more stable it will be, the opposite is in fact true. An increase in draft of about 1 foot can cause a considerable decrease in stability. This is why restrictions are set on how much deck load a unit may carry during ocean tows, in spite of the expense necessary for freighting consumables such as drill pipe and supplies out to the operating site.

Strength during Tow

Figure 6-7 illustrates the effect of roll on the stresses which occur in the legs of a jack-up during ocean tow. Similarly, the flexure of the structure of a semisubmersible cannot exceed the allowable stress values. Several accidents have occurred with jack-up units during tow in rough seas, in which the legs have been damaged or even lost. The units must therefore be designed to withstand these stresses. Based partly on theory and partly on practical experience, the criterion of a 20° off vertical roll at a 10 second period is used to determine the allowable length of leg that can rise above the upper leg guides during ocean tow.

Leg sections in many cases must therefore be removed prior to ocean tow in order to prevent damage to the legs. Similarly, if unexpectedly severe conditions occur during a field move, the legs may have to be lowered to avoid damage. Curves showing the amount of leg which must be removed (for ocean tow) or lowered (during field moves) are prepared as in Figure 6-7.

Strength on Location

A typical stress range curve for one location at a joint on a jack-up leg is shown in Figure 6-8. Two requirements must be

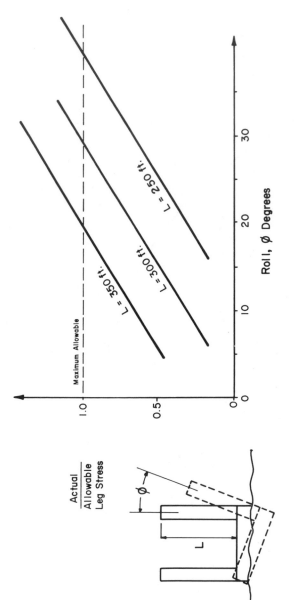

Fig. 6–7 Typical limitations on leg length for safety during tow.

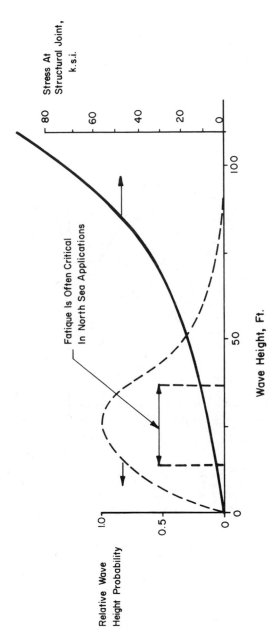

Fig. 6–8 *Typical effects of static and fatigue loading on structure of mobile offshore drilling units.*

satisfied for North Sea operation—static and fatigue criteria. Static criteria are well understood and have been used in the design of units for many parts of the world. The importance of fatigue criteria, however, has recently become evident in North Sea experience, as several fatigue failures have occurred in structural members. The frequency of occurrence of 10-25 foot waves which cause significant stresses in a structure is relatively high in the North Sea as compared with the Gulf Coast and most other offshore drilling areas.

This increased frequency makes fatigue much more of a problem even though the waves themselves are smaller than the maximum waves the unit was designed to withstand. For example, waves of 15-20 foot heights can be critical in causing fatigue in the North Sea because they occur so frequently, despite the fact that an 80-90 foot maximum wave occurring during a 50 or 100 year storm would be resisted.

Fatigue, rather than stresses encountered under the action of a maximum height wave (50 or 100 year storm), can sometimes therefore be the critical design case. This has caused increased attention to be focused recently on structural connection design and materials specification.

Strength for Different Criteria

Figure 6-9 shows typical curves for varying criteria. Rig owners and contractors have been using data such as these although classification and certifying authorities have not (to our knowledge) used them. However, several offshore mobile drilling units have been classed in recent years for more than one set of operating criteria. Certification for operation of an individual unit in new locations can be much quicker if the unit is classed for varying criteria or if different sets of criteria curves are available for that unit.

Bottom Soil Conditions

Bottom soil conditions are critically important for the safety of jack-ups as well as for the anchoring of mooring systems for drillships and semisubmersibles and the stability of column-

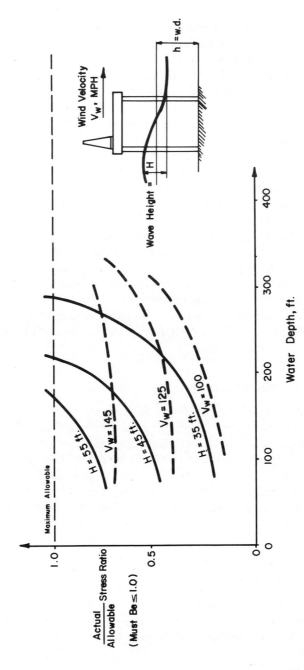

Fig. 6-9 *Typical curves for determining capability of jack-up unit.*

supported units and submersibles. For example, these conditions dictate the type of footing required to counteract the effects of leg penetration or scouring.

As an operationally important factor, these criteria have received more attention from certifying authorities than from the classification societies. However, as the classification societies take on more responsibility as consultants to the certifying authorities or as the designated certifying body of the governmental regulatory agency, their research into the effects of soil conditions and their requirements for the design of the entire unit considering these effects are increasing.

Figure 6-10 shows an example of a set of curves prepared for a series of mat-supported jack-up workover units operating in the Gulf of Mexico. Also shown is a set of curves for an independent leg (no mat) jack-up with the acceptable sea conditions for going on location in given bottom soil conditions.

WORKING WITH THE CLASSIFICATION AND CERTIFYING AUTHORITIES

It has been our experience in dealing with the classification societies that they are practical, experienced, and reasonable. They work closely with the rig owners and designers and are willing to consider the merit of any proposed exceptions or changes. However, they do require solid engineering back-up to justify any variances in policy they are asked to make. It is recommended that several steps be taken as early as possible in the planning of an offshore mobile drilling unit:

1. Get to know the representatives from the classification society or government authority as soon as possible. Let them see the care, thoroughness, and reasonableness of your approach to the use of the unit.
2. Start the classification and certification processes as soon as possible. There are two important reasons for this: the classification and certification authorities have a vast amount of work during the current boom in offshore rig construction and "rush jobs" or "last-minute surprises" may lead to disadvantageous delays. Design questions can

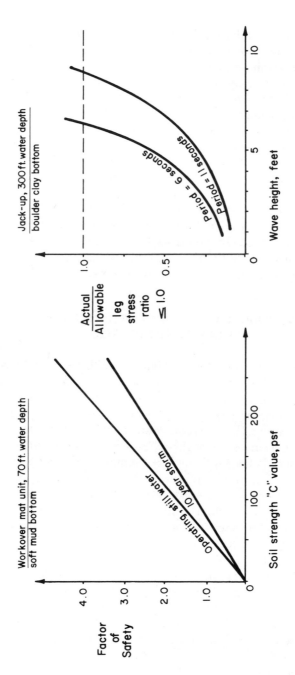

Fig. 6-10 *Effects of bottom soil conditions.*

be resolved before the construction of the unit is so far along that the required changes will cause costly and time-consuming corrections.

3. If you have any disagreement with a rule or procedure which the classification society or certifying authority is following, tell them. The offshore industry is so new that we can only learn from sharing the knowledge gained from past experience.

APPENDIX

Classification Societies

Name, Address	Major Publications, "Rules" Pertinent to Offshore Mobile Drilling Units
American Bureau of Shipping 45 Broad Street New York, N.Y. 10004 U.S.A.	"Rules for Building and Classing Offshore Mobile Drilling Units 1973" "Rules for Building and Classing Steel Vessels 1974"
Bureau Veritas 58, bis Rue Paul-Vaillaint-Couturier 92300 Levallois-Perrett, France	No publication applicable
Det Norske Veritas Grenseveien 92 Oslo 6, Norway	"Rules for the Construction and Classification of Mobile Offshore Units 1975" "Principles for Classification of Offshore Drilling Platforms 1973" "Rules for the Construction and Classification of Steel Ships 1970"

Name, Address	Major Publications, "Rules" Pertinent to Offshore Mobile Drilling Units
Germanischer Lloyd 1 Berlin 19 Heerstrasse 32 Germany	No publication applicable
Lloyd's Register of Shipping 71 Fenchurch Street London EC3M 4BS, U.K.	"Rules for the Construction and Classification of Mobile Offshore Units 1972" "Rules and Regulations for the Construction and Classification of Steel Ships 1974"
Nippon Kaiji Kyokai (Japanese Marine Corporation) 17-26 Akasaka 2-chome Minato-ku Tokyo 107, Japan	No publication applicable
Registro Italiano Vis Corsica 12 16128 Genoa, Italy	No publication applicable

Governmental Regulatory Agencies

Name, Address	Major Publications, "Rules" Pertinent to Offshore Mobile Drilling Units
Sjøfartsdirektoratet (Norwegian Maritime Directorate) P.O. Box 8123 Oslo 1, Norway	
U.S. Coast Guard Washington, D.C. 20591 U.S.A.	CG320 "Rules and Regulations for Artificial Islands and Fixed Structures on the Outer Continental Shelf, Subchapter N, July 1, 1972"
Occupational Safety and Health Administration U.S. Department of Labor Washington, D.C. 20210 U.S.A.	29CFR1910 "Occupational Safety and Health Regualtions, June 1974" 29CFR1926 "Occupational Safety and Health Regulations for Construction, June 1974"
U.K. Department of Energy (formerly Department of Trade and Industry) Petroleum Production Division Thames House South Millbank London SW1P 4QJ, U.K.	"Guidance on the Design and Construction of Offshore Installations 1974" "Offshore Installations (Construction and Survey) Regulations 1974" "Mineral Workings (Offshore Installations) Act 1971, Proposals for Construction and Survey Regulations," October 1972. "Mineral Workings (Offshore Installations) Act 1971, Environmental Factors Relating to the Design and Use of Installations," October 1972.

21

Marine Insurance

Frank J. Wetzel
Wetzel Surplus Lines, Inc.

Insurance and its relation to a company's operation is a complex subject. This section will deal primarily with the more specialized types of insurance for offshore drilling contractors operating in both national and international waters.

Insurance is largely separated into two main areas. First, there is protection against loss of life and personal injury to employees and members of the public and their property. Members of drilling crews operate complex equipment in all corners of the world, often under extremely hazardous conditions. This is not only true of the drillers' employees, but of all the specialized services that support the drillers: engineers, specialty tool people, divers, boat operators, helicopter pilots, etc.

When translated into dollars of potential liability the responsibility for the safety of these men and for environmental damage which might result from drilling operations represents staggering sums. Compensation rates alone for drillers' employees approach 40% to 50% of payroll costs.

The second area includes adequate protection against damage to or loss of drilling equipment. It is not uncommon to see some offshore drilling barge hull rates in the neighborhood of 7½ %, and this down from a high of 9¾ % just a few years ago. This equipment is so valuable on today's market that the loss of a drilling unit obviously can have a serious impact on the company's earnings.

So it is essential that a great amount of care and study be given to a proper insurance program. First, it is necessary to understand the character of the risk involved. The basic principle of insurance is that losses of the few are paid from premiums of the many. If a particular risk is too large to bear alone, then the decision is made to transfer that risk to the insurance company for a consideration, the premium. Of course, the degree of risk involved largely determines the premium to be charged.

In the offshore industry, however, the losses of the few are paid for by the premiums of the few. Although this situation is not as disproportionate as it was a few years ago, we still have a heavy concentration of values in fairly restricted geographical areas. The continued buildup in drilling equipment values in the Gulf of Mexico and the explosion of values in the North Sea present some extremely difficult problems apart from the routine hazards of drilling. Though wind has not caused the grief it did a few years ago, the possibility of its return to the forefront is strong. With its return, large monetary losses will be realized, particularly with the current inflationary costs.

Because losses have been less severe in the past several years, insurance rates have shown a steady decline. It is hoped that this will continue, but there are also two other factors which are contributing to the reduction in rates. The first of these is the improved design of the units, the techniques used, and the crews. The second is the worldwide deployment of drilling activity and the resultant spread of risk.

Of course, as the design, techniques and crews continue to improve, so too will the expansion of the areas in search for oil and gas. This will cause a continual pushing outward to new frontiers in which insurance underwriters will be expected to follow in this endless search. Extremes in weather and location accompanied with increased sophistication of equipment will keep pressure on rating higher than so called "standard" industry insurance costs. Thus far, there is a comparatively low number of contractors in this business, which also keeps rates high.

A brief word about the difference in rates for various designs. Generally, rates will be the highest for jack-up units, lower for semisubmersibles, and still lower for shipshape designs. Platform rigs, permanently fixed to the ocean floor, must be rated on the area located.

As more knowledge is gained about design and location, increased sophistication will be achieved in insurance rating. At one time, there was one rate for all designs irrespective of geographical location. This was because all units were jack-ups and almost all were in the Gulf of Mexico. Now, however, drilling operations are stretched from the North Sea to Indonesia, from offshore Chile to Sable Island, and all require special drilling conditions.

Who provides the insurance for these varied operations? The list is as varied as the operations, but generally it can be broken down into the same two categories as presented earlier. Workmen's Compensation and Employers' Liability and corresponding third party liabilities are written by a number of insurance companies, most of whom have little or no interest in writing the physical damage. For a number of years, physical damage coverage was insured by the London market. In recent years, the market has expanded as the size and number of units have increased. All of this completion is, of course, good for the industry because it brings competitive forces into the market.

Technology obviously has improved a thousandfold since the late 1940's when literally anything was put into the water just as long as it could hold a rig. Both drillers and underwriters cut their offshore teeth on such monstrosities. Unfortunately, many individuals lost their lives and suffered frightful monetary losses in the process of learning what type of equipment should be used and where.

Now it has become essential that marine architects and design engineers be brought in at the initial stages of conceptual design. Then, in addition, specialty firms should be employed by the owners to make certain that the proposed unit will meet insurance underwriters' standards both as to drilling capabilities and seaworthiness. All moves of these units must be accompanied by still another group of people, the marine surveyor.

In addition to attending the moves, the surveyor will perform a condition and value survey to establish the basis for insurance and, in fact, determine the method of loss adjustment to be used when and if a loss should occur. Thus, it becomes somewhat easier to assess a dollar loss when the assured and adjuster have long before determined what amount of insurance should be provided on the various components of these units.

Like any industry, the offshore drilling business has many complex features, some of which are just plain hard to solve. But at the same time, there is a continued willingness of the worldwide insurance industry to keep pace with the offshore drilling industry. As more capital is generated to drill deeper wells in more remote areas, the search for new capital is necessary in order to meet our requirements. Oil and gas are also becoming increasingly precious, as evidenced by the events of the recent past. It is absolutely essential that all involved answer the challenge. The insurance industry intends to meet that challenge.

The Authors

Ralph G. McTaggart

RALPH G. MCTAGGART

Ralph G. McTaggart is President of ETA Offshore Seminars, Inc. and is Vice President, Naval Architecture, of ETA Engineers, Inc. in Houston, Texas. He has been working in the offshore industry since 1954 and has been involved in the design, analysis, certification, and construction of offshore drilling and production units. Mr. McTaggart is on the Special Committee on Offshore Mobile Drilling Units of the American Bureau of Shipping and is a member of the Society of Naval Architects and Marine Engineers and the Institute of Marine Engineers. He received his degree in Naval Architecture from Stow College in Glasgow, Scotland, is a Chartered Engineer (U.K.), and holds a City and Guilds of London Certificate in Shipbuilding.

Odd A. Olsen

ODD A. OLSEN

Odd A. Olsen is a Research Engineer with the Division for Marine Technology of Det norske Veritas (DnV) in Norway. With DnV, he is primarily responsible for the study of wave loads on offshore structures. Mr. Olsen has served as a Scientific Assistant in the Department of Ship Structures at The Technical University of Norway in Trondheim. He received his Bachelor of Science degree with a specialty in structural problems in container ship design from the University of Norway and completed the studies for Naval Architect.

Mark A. Childers

MARK A. CHILDERS

Mark A. Childers is Supervising Engineer with Ocean Drilling and Exploration Company (Odeco) in New Orleans, Louisiana. He has worked in many phases of production and drilling and has specialized in the development of subsea drilling equipment and the design and analysis of floating drilling units. Mr. Childers is a member of the American Petroleum Institute's Task Group on Mooring Gear for Floating Structures. He holds several patents, is a registered Professional Engineer in Texas and Louisiana, and in 1969 he received the SPE AIME Cedric K. Ferguson Medal. Mr. Childers received his Bachelor and Master of Science degrees in Civil Engineering from Virginia Polytechnic Institute and State University.

Tom R. Reinhart

TOM R. REINHART

Tom R. Reinhart is Manager of Research and Development for the Baylor Company in Sugarland, Texas, where he is involved with offshore instrumentation projects, oceanographic systems, variable voltage power systems, ballast control systems, and data gathering and reduction processes. He has worked with dynamic positioning on the vessel "Eureka" and served as Engineering Manager for the work on "Sedco 445." Mr. Reinhart holds patents for electrical, mechanical, and instrumentation systems. He received his Bachelor of Science degree in Mechanical Engineering from MIT and his Masters degree in Mechanical Engineering, with a minor in Electrical Engineering, from the University of Houston.

William L. Clark

WILLIAM L. CLARK

William L. Clark is Product Manager of Drilling Tools for Hydril Company in Houston and is responsible for the market application of existing and new Hydril drilling tools and new product development. He has over ten years of experience in the application and design of drilling equipment. Mr. Clark has also served as an Instrumentation and Control Systems Engineer in the research laboratories at Stanford University. He is a member of SPE, AIME, and the Nomads and is a lecturer for training seminars for Hydril, ETA Offshore Seminars, Inc., and the University of Texas. He received his Bachelor of Science degree in Industrial Engineering from Georgia Tech and is a registered Professional Engineer in Texas.

D. Bynum, Jr.

DR. DOUGLAS BYNUM, JR.

Dr. Douglas Bynum, Jr. is Manager of Research Engineering for ETA Engineers, Inc. in Houston. He is responsible for the technical studies and business development of marine pipelay operations, advanced types of offshore drilling units, and special structural analyses. Dr. Bynum has fifteen years of experience in machine design, structural analysis and design, corporate planning, and management. He is a registered Professional Engineer in Texas and is a member of the Society for Experimental Stress Analysis, Chi Epsilon, and Sigma Xi. He received his Bachelor of Science degree in Mechanical Engineering from Texas A&M University, his M.S. degree in Mechanical Engineering from the University of Washington, and his Ph.D. in Interdisciplinary Engineering from Texas A&M.

R. Maini

RAMESH MAINI

Ramesh K. Maini is a Structural Engineer with ETA Engineers, Inc. in Houston where he is involved with the design and analysis of offshore drilling units. He has been responsible for the structural arrangement and design of jacking systems and spud tanks for jack-up rigs. Mr. Maini has also worked with design and analysis of pipe bending shoes and iceberg loading and icebreaker barge analysis. He is a member of the Houston Engineering and Scientific Society and an associate member of the American Society of Civil Engineers. He received his B. Tech degree in Civil Engineering from the Indian Institute of Technology and his Master's degree in Civil Engineering from the University of British Columbia.

Mike Reifel

DR. MIKE D. REIFEL

Dr. Mike Reifel is an Advanced Senior Petroleum Engineer in the Offshore Technology Division of Marathon Oil Company in Houston. He has conducted research in advanced methods to analyze structural and fluid systems and has managed numerous offshore pipeline and platform projects for the oil industry. Dr. Reifel has jointly developed a method to kill offshore well blowouts with a major oil company. He is a registered Professional Engineer in Texas. He received his Bachelor of Science degree in Mechanical Engineering, his M.S. degree in Engineering Mechanics and his Ph.D. from the University of Texas in Austin.

Hugh L. Elkins

HUGH L. ELKINS

Hugh L. Elkins is Director of Marketing—Mechanical Products for Hydril Company in Houston and is responsible for overall product management. He has worked with the supervision of product development for subsea pipeline equipment and other marine equipment applications. Mr. Elkins has managed the equipment manufacturing and production phases. He is a member of IADC and Nomads and has been an instructor and author of material for Hydril, IADC, ETA Offshore Seminars, Inc., and University of Texas programs. He studied Petroleum Engineering at the University of Houston.

John McLain

JOHN D. MCLAIN

John D. McLain is the Worldwide Subsea Drilling Sales Manager for the Stewart and Stevenson Oilfield Division in Houston. He has had experience as a Project Leader for research on subsea drilling equipment and has been involved in onsite surveys of drilling rigs off Australia and in the North Sea. Mr. McLain has served as drilling superintendent for all types of land and offshore rigs. He has also supervised the engineering design for artificial lift workover and drilling equipment. He received his Bachelor of Science degree in Chemical Engineering from Texas A&M University.

Darrell Foreman

DARRELL G. FOREMAN

Darrell G. Foreman serves as the Worldwide Sales Manager for Subsea Equipment with the Stewart and Stevenson Oilfield Division in Houston. He worked on the world's first deep water unit operating in 1,400 foot depths off California. Mr. Foreman also served as the Project Manager for the first acoustic blowout preventer system.

W. B. Bleakley

W. B. BLEAKLEY

W. B. Bleakley is the Marketing Supervisor of the U.S. region for Lockheed Petroleum Services, Ltd. in Houston where he works with the company's one-atmosphere, manned, subsea oil production systems. He has served as a reservoir engineer and was on the petroleum engineering faculty at the University of Tulsa. Mr. Bleakley was also Production Editor for *Oil and Gas Journal* for sixteen years. He is a member of SPE and is a registered Professional Engineer in Oklahoma, Texas, Nebraska, and Wisconsin. He received his Bachelor and Master of Science degrees in Petroleum Engineering from the University of Tulsa.

Mike Hughes

MICHAEL D. HUGHES

Michael D. Hughes is Chairman of the Board of Oceaneering International, Inc. in Houston. He directs a group of companies which provide diving services to the oil industry, including pollution control and a commercial diving school. He assembled and introduced the first underwater television-video system for subsea inspection and assisted in the design and construction of equipment for the first offshore mixed-gas dives. Mr. Hughes is a registered Professional Engineer in Texas and Louisiana and is a member of the Association of Diving Contractors, Marine Technology Society, National Ocean Industries Association, and others. He received his Bachelor of Science degree in Civil Engineering from the University of Tennessee.

Larry Cushman

LARRY CUSHMAN

Larry Cushman is Special Projects Manager for Oceaneering International, Inc. in Houston. He is the program manager in charge of the construction of the Oceaneer 1000, the first 1000-foot saturation diving system designed for use aboard a semisubmersible drilling vessel. He is President of Oceaneering's diver training school, the Commercial.Diving Center in Los Angeles, and served as director of the school for two years before coming to Houston.

Prior to joining Oceaneering, Mr. Cushman was with Sub-Marine Systems, an R & D company in Los Angeles engaged in the development of cryogenic closed-circuit mixed-gas SCUBA, cryogenic chamber life support systems and chemical diver heaters. He was one of the founders of the company and worked. as a project manager, test diver and underwater photographer. Mr Cushman has been professionally involved in the diving business as a manager, instructor and commercial diver since 1961. He is currently Vice President of the National Association of Underwater Instructors.

J.E. Weatherly

J. E. WEATHERLY, JR.

J. E. Weatherly is President of Weatherly Engineering, Inc., Weatherly Laboratories, Weatherly International, and Pollution Control International in Victoria, Texas. He supervises the companies which are active in production testing, hydrocarbon analyses, PVT studies, and super oil burners for pollution control for land and marine operations. Mr. Weatherly has served as a drilling, production, and Petroleum Engineer at Texas A&M University. He is a member of API, AIME, and SPE. He received his Bachelor of Science degree in Petroleum Engineering from Texas A&M University.

Peter M. Lovie

PETER M. LOVIE

Peter M. Lovie is President and co-founder of Engineering Technology Analysts, Inc. and is Senior Vice President of ETA Engineers, Inc. in Houston. He has experience in the study of strength and behavior of offshore drilling units under storm conditions and during tows, and is now responsible for client liaison and new business development. Mr. Lovie is a registered Professional Engineer in Texas, a Chartered Engineer, and a Chartered Naval Architect (U.K.) He is a member of ASME, RINA, ASCE, and SNAME. Mr. Lovie received his Bachelor of Science degree from the University of Glasgow, Scotland, and his Master of Applied Mechanics degree from the University of Virginia.

Frank J. Wetzel

FRANK J. WETZEL

Frank J. Wetzel is the President of Wetzel Surplus Lines, Inc. and specializes in oil insurance and commercial marine insurance. He has been involved with insurance for drilling contractors since 1952. Mr. Wetzel attended the University of Texas and the University of Houston.